PREFACE

The Thrust of modern scientific effort is reductionism - breaking parts into smaller fractions in search of an entity from which all else depends.... in the end, there is no substance, no geometry, no particles nor even a wave. Only relational connections are revealed. The universe is a jig-saw puzzle, understood by putting pieces together. It cannot be synthesized by dissection.

THE COVER

The Celestial Rose formed by the interaction of two galaxies may no longer be in bloom. Light now arriving at the Hubble telescope has been traveling for more than 300 million years.

"At the heart of everything is a question, not an answer. When we peer down into the deepest recesses of matter, or at the farthest edge of the universe, we see, finally, our own puzzled faces looking back at us."

John Archibald Wheeler

FIELD FORCES FROM FIRST PRINCIPLES

Dedicated to the Memory of my Mother Carol and my Uncle Roy

A Note of Thanks for the Continued Support of:

Robert and Gloria Potter

James J Brennan

And My Dear Wife,

Astrid

FORWARD

The means by which material objects extend their presence across the void is yet a mystery. Newton considered space as absolute, a stage upon which the particles performed. Einstein believed all motion was relative, mass conditioned space,

The discovery of cosmic background radiation (CBR) in 1966, re-enforced the idea of a genesis. In the years following, the exponentially decelerating universe became the defacto standard. The standard model was a big bang followed by some 14 billion years of gravitational deceleration. That all changed in 1998 with the Lawrence Berkeley **1a** supernova studies. The theory of the universe was turned upside down. For the author, accelerating expansion came as welcome news after years of frustration attempting to model gravity from a flawed premise. Within these pages, the reader will find the rest of the story.

In giving audience to a young physicist seeking advice, Einstein counseled it would be better to earn a living in some unrelated labor that did not involve constant academic scrutiny, or as he put it: "Get a cobbler's job that you may have the liberty to ponder your ideas and make mistakes in private." So it is with this Thesis. The author has made many mistakes in private, and now offers these pages to the reproval of its critics.

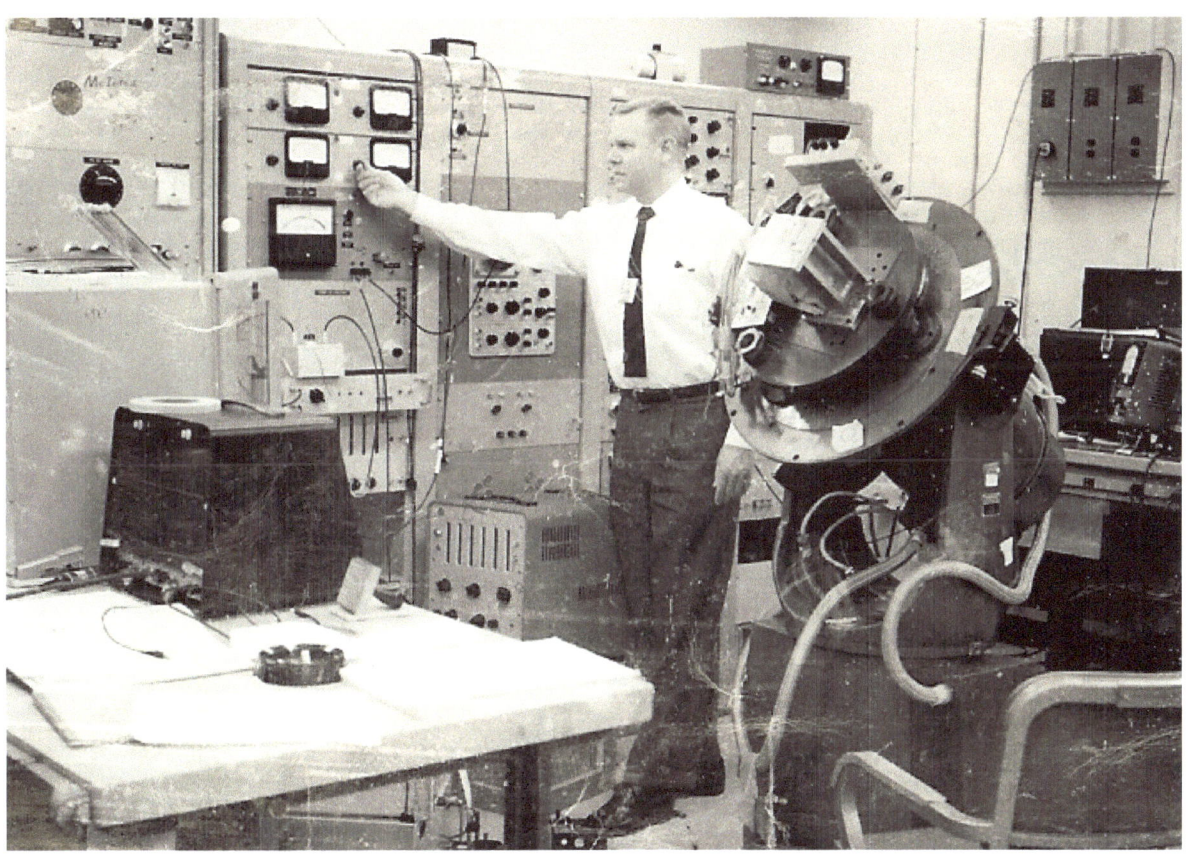

The Author, TRW Inertial Guidance Lab, Circa 1965

TABLE OF CONTENTS

Part I

Feynman's Gravity - Folly or Foresight 10

Gravity From Expansion 12

The Variable Constants 26

Part II

Prelude to Charge 49

Electric Particles as Expanding Angular Momentum 51

PART I

Why are the equations that describe such different physical phenomena so similar? We might say: It is the underlying unity of nature. But what does that mean? What one thing is there that is common?" Nobel Laureate Richard Feynman answered his own rhetorical: "....it is the *space*, the framework into which the physics is put."

FEYNMAN'S GRAVITY - FOLLY OR FORESIGHT

"One very important feature of pseudo forces is that they are always proportional to the masses. The same is true of gravity. The possibility exists therefore that gravity itself is a pseudo force. Is it not possible that perhaps gravitation is due simply to the fact we do not have the right coordinate system?"

Feynman - Lectures on physics

For Richard Feynman, the idea that gravity could be a delusory manifest of some known phenomena seemed to be always with him. In his Lectures on Gravity, he frames the issue thus:[1]

1) Gravitation is a new Field of its own, unlike anything else, or

2) Gravity is a consequence of something already known but incorrectly perceived.

New Physics requires strong evidence. The empirical support for a theory of gravitons congruent with the successful predictions of Einstein's geometric was then, and is today, still missing. But does static space really curve, and if not, do gravitons really exist? If neither, then what slumbers within the tiniest increments of matter to summon one mass to another. Perhaps the culprit is space itself. Indeed, some characteristics commonly attributed to an empty volume are similar to those of solids, liquids and gases. Still, well understood descriptors such as expansion, distortion, and pressure, can be nebulous and even contrary when applied to a massless void.

Feynman referred to forces that result from acceleration as *"pseudo forces"* (instantaneous inertial opposition proportional to mass).[2] The identity of the structure undergoing acceleration, however, is not specified by Newton's second law, at least not by Newton. Symmetry of action, however, is a mandate of conservation, in the case of gravity and inertia, it is conservation of force. As Einstein foretold, inertial reaction and gravity are opposite sides of the same coin. If local '**g**' fields are negative pressure gradients, how is virtual stress stored? Dimensionally, pressure is momentum flow, and for the cosmos, momentum flow is spatial expansion.

Reversing the roles between space and matter excites new possibilities. While working out the theory of General Relativity, Einstein explored the properties the universe must posses to prevent the detection of absolute motion.[3] What followed was the principle of relative acceleration...the force felt by the crew of an accelerating rocket ship is no different than that experienced by the same crew at rest in a universe undergoing unidirectional acceleration. How might this pseudo force be distinguished? Although not recognized as such, the answer would come with the discovery of cosmological expansion. Accelerating objects feel inertial reactions as pseudo forces. Masses subjected to isotropic spatial acceleration, create counter reactions. Embellishing upon Feynman's musings, we look further into the dimensionality of G and relationship to global expansion.

---

[1]Feynman, Lectures on Gravity, Lecture 1, §1.5

[2]Feynman, Lectures on Physics, Vol I, §12-5

[3]'Understanding Physics, Isaac Asimov, Barns and Noble 1993. Mechanism at pages 115-120:

Einstein perceived gravitational and inertial mass as equivalent. Since there was no separate and identifiable "gravitational mass" *per se*, it was necessary to postulate a property of space that could be affected by the presence of matter. To enable inertial mass to bring about a physical result, static space needed to curve in the proximity of condensed energy. Einstein solved the requirement by turning the problem into a postulate which has become a principle. While the idea of distorted space is plausible, the functional mechanism remains unknown. What properties must space possess that would cause it to curve or otherwise exhibit unusual behavior in the presence of mass. If inert matter conditions space, how does space affect matter?

As developed herein, inertial matter at rest in expanding space feels reactive forces analogous to those measured when objects are accelerated relative to a rest frame. Because matter is composed of particles held together by electric and quantum forces, these forms of energy are not disassociated or deformed by the relatively weak expansion field. Expansion resistant masses create counter fields proportionate to their masses. To complete the formalization of gravity as a pseudo force within Feynman's denotation, a corporal vitality must be identified and quantified.

"What I cannot create, I do not understand"

**Written by Richard Feynman in the corner of his office blackboard
Where it remained for eight years**

GRAVITY FROM EXPANSION

The Effect of Spatial Expansion upon Non-Expanding Matter

How does space communicate opposition to changing velocity? Historically, instantaneous reaction has been the Achilles heel of Machian Mechanics, and most likely the reason for Einstein's reluctance to embrace it as a perspective of General Relativity. The average density of the universe is immeasurably small, offering nil resistance to kinetic change. Yet in some manner, counter forces appear instantly when masses and charges are accelerated. Is a similar reactionary effect observed when isotropic expansion acts upon matter? For empty space to be a player in the gravitational scheme, it must function as a dynamic.

Newton's 2nd law equates inertia to rate of change of momentum. For non-expanding matter, the relationship between force, mass and space can be written in terms of pressure $\mathbf{P}$ as:

$$\frac{\mathbf{F}}{\mathbf{A}} = \left[\frac{\mathbf{M}}{\mathbf{A}}\right]\mathbf{a} = \mathbf{P} = \sigma(\mathbf{a}) \tag{1}$$

where both force and mass have been divided by an area 'A.' In this new cast, exposition, of gravitational action proceeds from a simple dimensional reformulation. Specifically, *global volumetric acceleration per unit mass* expressed as *directional acceleration moderated by scalar density*. The dimensionality of Newton's mysterious $\mathbf{G}$ factor provides a clue to its origin:

$$\mathbf{G} = \left[\frac{\mathbf{meters}^3}{\mathbf{second}^2}\right] \mathbf{x} \left[\frac{1}{\mathbf{kgm}}\right] \mathbf{x} \left[6.67 \times 10^{-11}\right] \tag{2}$$

$\mathbf{G}$ is thus volumetric acceleration per unit mass per (2) or unidirectional acceleration imposed by a surface density per (3)

$$\mathbf{G} = \left[\frac{\mathbf{meters}}{\mathbf{second}^2}\right] \mathbf{x} \left[\frac{\mathbf{meters}^2}{\mathbf{kgm}}\right] \mathbf{x} \left[6.67 \times 10^{-11}\right] \tag{3}$$

While (2) and (3) are dimensionally identical, they express different conceptual notions. Thus, if the scale of the Hubble sphere is '$\mathbf{R}$' and $\mathbf{M_u}$ represents what is called '*bare mass*' defined as the sum of all individual masses absent any gravitational interaction (e.g., that obtained by separating matter into infinitesimally small units spaced far apart). The gravitational energy for the 3-sphere volume of the Hubble sphere is:

$$\mathbf{U_3} = -\frac{3\mathbf{M_u}^2\mathbf{G}}{5\mathbf{R_3}} \tag{4}$$

Negative gravitational energy expressed per (4) is tagged with different names (as *mass-defect*, *binding-energy*, or *gravitational deficit*). It accounts for the work-energy required to distribute $\mathbf{M_u}$ uniformly throughout the Hubble volume working against the gravitational field.

If instead, $\mathbf{M_u}$ is spread uniformly over the surface of an imaginary Hubble manifold creating a 2-sphere, binding energy is reduced to:

$$U_2 = -\frac{M_u^2 G}{2R_2} \tag{5}$$

Binding energy adds to bare mass energy, but for a self creating adiabatic universe to evolve *ex nihilo*, total energy must equal zero. The effective Hubble mass [bare energy plus negative binding energy] must be twice bare energy. Calling $\mathbf{Mc^2}$ energy positive, negative energy in the form of gravitational fields will contribute an equal amount to the total energy $\mathbf{M_u}$, hence (5) leads to the following:

$$M_u c^2/2 - M_u^2 G/2R_2 = 0 \tag{6}$$

Whence

$$\frac{M_u G}{R_2 c^2} = 1 \tag{7}$$

And for the 3-sphere from (4):

$$M_u c^2/2 = 3M_u^2 G/5R_3 \tag{8}$$

Therefore

$$\frac{M_u G}{R_3 c^2} = \frac{5}{6} \tag{9}$$

The difference between $\mathbf{U_2}$ and $\mathbf{U_3}$, arises from difference in the gravitational binding energy incurred when transforming from 3-sphere to 2-sphere. For two geometries to have the same energy requires $\mathbf{R_2 = [5/6]R_3}$. For purposes of numerical computation, a plausible "standard model" value for the Hubble radius $\mathbf{R}$ will be taken as $\mathbf{1.3 \times 10^{26}}$ meters. The trial value of $\mathbf{M_u}$ gathered from an average of several independent methods is provisionally $\mathbf{1.5 \times 10^{53}}$ Kgm.[4] When the 2-sphere radius is appropriately reduced to account the difference in the gravitational deficit, the 2-sphere (7) and 3-sphere (9) formulations lead to the same value for $\mathbf{G}$:

$$G = \frac{R_2 c^2}{M_u} = \frac{5}{6}\frac{R_3 c^2}{M_u} = \frac{5}{6}\left[\frac{(1.3 \times 10^{26})(9 \times 10^{16})}{1.5 \times 10^{53}}\right] = 6.5 \times 10^{-11} \frac{\text{meters}^3}{\text{sec}^2 \text{kgm}} \tag{10}$$

The value of $\mathbf{G}$ expressed in (10) is based solely upon energy considerations. As anticipated from its dimensional structure, $\mathbf{G}$ boils down to Hubble factors, [*global volumetric acceleration divided by total mass-energy*]. In terms of 3-D density and volume, (9) can be expressed as $\mathbf{15H^2/24\rho_U\pi}$.

---

[4]*Estimations of Total Mass and Energy of the Universe*, Dimitar Valev. arXiv:1004.1035v1 [Physics.gen-ph] 4-7-2010. Leads to same dependence of $\mathbf{G}$ upon $\mathbf{R}$. (Changes in $\mathbf{M_u}$ should be a function of $\mathbf{R^2}$ rather than $\mathbf{R}$). See also: *Derivation of Mass of the Observable Universe.* Carvalho J.C. Int. J.Theor. Phys., 34, 1995, 2507-2509

The **5/6** adjustment follows from the fact that volumetric density ρ_U requires R_3 to be cubed leaving R_3^1 hidden in the numerator when **G** is expressed in terms of **H**:

$$G = \frac{5}{6}\frac{3H^2}{4\pi\rho_u} = \frac{5}{6} \times \frac{\dfrac{c^2}{R^2}}{\dfrac{M_u}{R^3}} = \frac{c^2 R_3}{M_u} \times \frac{5}{6} \tag{11}$$

whence (11) differs from previous attempts to express **G** in terms of Hubble parameters due to the inclusion of the 3-sphere gravitational deficit. For the 2-sphere construct, R_2 is required to have a reduced radius $= \mathbf{1.083 \times 10^{26}}$ meters if M_u and **G** remain unchanged:. Hence:

$$G = \frac{H^2}{4\pi(\sigma_u R_2)} \tag{12}$$

But (12) is simply a rearranged expression for Friedmann's formulation when the universe is undergoing de Sitter expansion. That is, when $q = -1$, expansion can be expressed as c^2/R whence:[5]

$$\rho_u = -q\frac{3H^2}{4\pi G} \tag{13}$$

When (13) is transformed from a volume density to a surface density, it will be recognized as the numerical form of the dimensional expression (3). Substituting from (17) below, Newton's second law can thus be understood in terms of a universal scalar density function σ_u. As to be developed, σ_u is the connective between inertia and gravity for flat space. The 1st step is transforming Hubble mass M_u to a 2-sphere shell, whereby the radius is reduced so that total energy remains the same. The adjusted radius R_2 $(\mathbf{1.083 \times 10^{26}}$ meters) is based upon an estimate of the real 3-sphere Hubble radius $(\mathbf{1.3 \times 10^{26}}$ **meters**). The surface density σ_u is therefore:

$$\sigma_u = \frac{M_u}{4\pi R^2} = \frac{1.5 \times 10^{53}\,\text{kgm}}{4\pi[(1.083) \times 10^{26}\,\text{meters}]^2} \approx \text{one}\ \frac{\text{kgm}}{\text{meter}^2} \tag{14}$$

When the whole of mass contained within the Hubble volume is spread over the Hubble manifold, volumetic density ρ_u is replaced by a surface density $3\sigma_u/R_2$ so (13) becomes:

$$G = -\frac{q}{\sigma_u}\frac{c^2}{4\pi R_2} \tag{15}$$

For de Sitter expansion, $q = -1$, hence:

$$G = \frac{c^2}{4\pi R_2 \sigma_u} \tag{16}$$

---

[5]The symbol "**q**" is a dimension-less parameter formed by the ratio of 3 Hubble factors i.e., acceleration, velocity and scale: $-\mathbf{q} = \ddot{R}R/(\dot{R})^2$

The component parts of (3) correspondence to those of (16), that is $1/\sigma_u$ = **one [meters²/kgm]** and for **q = -1**, the acceleration factor is **c²/R₂**   As a conceptual vehicle, the 2-sphere universe provides a perspective of gravity and inertia as a unified functionality. This bounty of cosmological action within the economy of a single cause has been given the name "*inertio-gravitational action*."[6] The scalar density function [sigma], being numerically near one kgm per square meter, portends the intriguing proposition that it is exactly "one." Not by coincidence, but because that is the way it must be..... to fit the dimensional values established by the 17th Century experimenters in their efforts to relate *force*, *mass* and *acceleration*. Little could they have known the measured reactionary forces encoded both the size and mass of the universe. That local forces are determined by global parameters should come as no surprise to the reader.[7] Why inertia is defined by Hubble parameters, to the exclusion of what lies beyond, poses another question.[8]

As recapped by the red arrow in **Figure I-A**, a uniform 3-sphere transforms to a 2-sphere having the same external gravitational characteristics if the radius **R₃** is reduced to account for the difference in binding energy. The blue arrow depicts a further transformation from 2-sphere to flat surface 'S', the details of which are covered in Appendix __. What is significant for 'S' is that it can be treated as an infinite plane with respect to a mass 'E' small in area compared to the 2-sphere surface. The action of the infinite plane acting upon **E**, whether inertial or gravitational, will be independent of its distance from the plane. Newton's 2nd Law can now be understood in terms of a scalar density function **σ_u**.   That this field has a value of *one-kgm-per-square-meter* will complete the derivation of **G** from expanding space.

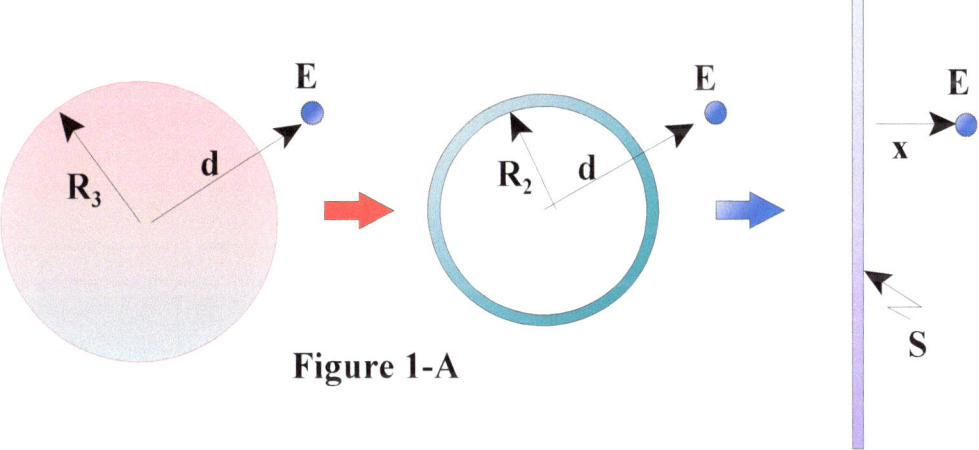

Figure 1-A

[6]The name "*inertio-gravitational field*" was first used by John Stachel to clarify relativistic unification(s).

[7]To assign a value to sigma in (14), there is much to suggests the naturally occurring relational ratio of Hubble mass **M_u** divided by Hubble surface area **4π(R₂)²**. Selected parameters for **M_u (1.5 x 10⁵³ kgm)** and radial scale **R = (1.3 x 10²⁶ meters)** transform to a one **kgm/m²** 2-sphere surface density. Reactionary force always equals accelerating force, and as to be shown, reactionary force per square meter corresponds to gravitational force, which is known with reasonable precision for some celestial objects.

[8]For most purposes, any size sphere will do if it is sufficiently large to include enough mass to determine average cosmic density. The Hubble is convenient because the radius is determined by '**c**.' The sample space must also be large enough to function as an infinite plane when matter content is transformed to a flat surface.

The solid sphere and the shell project the same gravitational force of at distance 'd' (as measured from the center of either), the lost gravitation deficit being recovered by reducing the shell radius R_2 to $(5/6)R_3$. Conversely, if R_2 is increased indefinitely, the shell area $(4\pi R_2^2)$ can be modeled as an infinite plane 'S,' thence, the gravitational deficit is reduced to zero. The interaction between **E** and **S** will be independent of the distance '**x**,' and (per 5) the density of the surface σ_S will be reduced by ½ because infinite radius reduces binding energy to zero.

While farther away generally means weaker force, increasing the distance '**x**' between **E** and **S** results in more mass acting at a less obtuse angle, so the intensity of the gravitational '**g**' field is the same at any distance '**x**'[9] The Hubble sphere centered on the observer nonetheless, continues in its function as the sample volume for determining average density, acceleration, and reactance when the universe is modeled as a flat plane. Since the same quantity of mass-energy is operative in the shell model, then

$$\rho_u(4/3)\pi R^3 = \sigma_u(4\pi R^2)$$

Hence:

$$\sigma_U = \frac{\rho_U R}{3} \tag{17}$$

Likewise, the result obtained by integrating over volume will correspond to the flux exiting across the surface *a la* the divergence theorem. Specifically,[10]

$$\iiint_V \rho_u\,\mathbf{dV} = \iint_S \sigma(\mathbf{dA}) \tag{18}$$

The results of the many CBR experiments evidence the likelihood that space is geometrically flat on the large scale. This raises the question of whether transformation from 2-sphere to flat plane differs from the 2-sphere model where $R \to \infty$. From (5), gravitational deficit $\to 0$ as $R \to \infty$. For any three dimensional volume the **g** field from Gauss's law is:

$$\int_S \mathbf{g}\cdot\mathbf{n(dA)} = 4\pi\mathbf{Gm}$$

then

$$\frac{\mathbf{g[A]}}{\mathbf{A}} = \frac{4\pi\mathbf{Gm}}{\mathbf{A}} = 4\pi\mathbf{G}\sigma \tag{19}$$

---

[9]The infinite flat plane is also referred to as a "Bouguer plate." For any finite thickness, the gravitational field is perpendicular to the plate with magnitude $2\pi G[\sigma_P]$ where σ_P is the mass per unit area of the plate.

[10]When transforming between volumetric density ρ_U and surface density σ_U for entities such as individual stars and planets, gravitational energy is insignificant compared to bare mass. Consequently no adjustment is required to account for the difference in binding energy when making a volume to surface transformation. For the universe, gravitational energy equals bare mass, ergo, correction is required to preserve the field strength.

As an aside, (19) leads to the same expression for **G** as will be later developed from spatial expansion. Specifically, if a Gaussian surround is applied to the present Hubble sphere then 'g' can be analogized to the accepted cosmological acceleration c^2/R apropos of **q = -1** de Sitter expansion:

$$\mathbf{G} = \frac{\mathbf{g}}{4\pi\sigma} = \frac{\dfrac{c^2}{R}}{4\pi\sigma} = \frac{c^2}{4\pi R\sigma} = \frac{c^2}{4\pi R_2\sigma} \tag{20}$$

Returning to the infinite plane, of area 'A' used to for '**σ**' is double, i.e., there are two sides, so the strength of the **g** flux is halved when integration is taken over the surface of both sides. Since all force lines are normal to the plane, only those on one side can act upon distant objects. Consequently **g = 2πGσ**. Thus in both the infinite 2-sphere and Euclidean space, half the energy is lost or otherwise unavailable to augment gravitational action. In both cases, however, the prevailing presumption holds, namely that negative gravitational energy must equal positive mass energy. For the expanding 2-sphere, this condition can be satisfied if negative gravitational energy increases with radius. For expanding flat space, the Hubble 2-sphere can still serve as the measuring volume, and as such, the ratio of positive mass energy to negative gravitational energy always = **1**. That energy loss is the same whether by transformation to an infinite plane, or expanded to R –> ∞, leads to the following Lemma: If both **G** (equation 16) and negative gravitational energy **U₂** (equation 5) are inversely proportional to **R**, then inertial mass will increase proportionately with Hubble area. Specifically from (7) and (16):

$$\mathbf{M_U} = \frac{\dfrac{R_2 c^2}{1}}{\dfrac{c^2}{4\pi R_2\sigma}} = 4\pi R^2\sigma \tag{21}$$

Which is the same as (14). As a check, the effective radius $R_2 = (1.083 \times 10^{26}$ **meters**$)$ corresponds to a Hubble sphere transformed to 2 sphere surface density, whence:

$$\mathbf{M_u = (12.56)(1.083 \times 10^{26})^2 \approx 1.5 \times 10^{53} \ kgm} \tag{22}$$

If positive matter energy equals negative gravitational energy in a flat space universe, the distinction between the mechanics of flat plane and the two sphere, although based upon different geometries, can be disregarded - the totality of energy in a representative volume of flat space is taken as it is found and the equivalent sigma plane is calculated from (21). For flat space, the infinite plane is an extension of the 2-sphere where, in the limit **R --> ∞**. No further transformation is needed to bring about the scalar-density plane as a functional operative. As a reactive element, σ_U defines inertial opposition in terms of a universal scalar surface density (per Newton's 2nd Law). Treating the scalar density field as an intrinsic characteristic of a flat space, our objective reduces to that determining a value that explains gravity in terms of inertia and vice versa. One inertio-gravitational field serves all occasions.

The inertial force per (1) for a particle M_E accelerated against a pressure 'P' is the integral of P over the area 'A' of its surface. Thus if P is the cosmological pressure and σ_U one **ntn/m²**, and 'a_n' equals the expansion rate then

$$\sigma_U = P/a_n \tag{23}$$

To accredit the assumed one **kgm/meter²** scalar density function as the essence of a common inertio-gravitational field, the earth will be used as a gravitational test particle. Invoking Newton's 3rd Law as the operative between earth and universe, then the gravitational field of the earth acting upon the universe equals the cosmological expansion field acting upon the earth. This proposition will be explored now and later from both a gravitational and inertial perspective. Introduced here as evidentiary support for both the formulation of the cosmological source field as expansion created spatial acceleration and the earth's '**g**' field as the inertial thereof. A secondary objective will be that of exploring the scalar density field as the means by which the all other mass in the universe creeps into the expression of local inertial reactance.

Taking the primary force as that imposed by the acceleration field created by the expansion of space as a_n, the for the mass of the Hubble universe transformed to an infinite flat plane (or a number infinite flat planes (each of infinitesimal thickness) presenting a total energy density in any and all directions σ_U, we arrive at an isotopic energy density surface. But contrary to current thinking no field or flux is created by the plane itself. The global G field of a plane or system of planes, arises when the density function is accelerated. Because expansion is isotropic, so also is the global G field. The intensity of the earths reactionary '**g**' field will be equal to the mass of the earth M_E divided by the area of the earth A_E multiplied by the spatial acceleration factor a_n. This corresponds to a dimensional ratio **ntn/m²**. That momentum be conserved on the global scale, the reactionary field created by the earth's resistance to expansion must cancel the pressure created by the action of the expansion field, specifically the reactionary field of the earth must act upon the universe, ergo:

$$\frac{M_E a_n}{A_E} = g\frac{M_u}{A_u} \tag{24}$$

Restated with 'E' subscripted as 'e,' the inertial affect of matter upon expanding space follows from Newton's 2nd law. The isotropic acceleration field acting upon M_e creates counter reaction $M_e(a_n)$. In treating M_e as a spherically uniform inertial mass of radius R_e, the counter pressure at the surface will equal the source pressure that produced the counter action:

$$\frac{F}{A} = \frac{M_u}{A_u}(g) = \frac{M_e}{A_e}(a_n) = \sigma_u g \tag{25}$$

Equation (25) is Newton's 2nd law in gravitational field form: "*The pressure created by the cosmological acceleration field a_n acting upon a spherically uniform density having mass to area ratio M_e/A_e is equal to the pressure created by the 'g' field counter acceleration acting upon the cosmological scalar density function σ_u.*"

From a Gaussian perspective, the flux exiting across earth's surface balances the flux emanating from the Hubble surface. But in contrast to the usual gaussian surface, A_e must be concentric with a uniform spherical mass. A_e is, however, conveniently enlarge-able to facilitate measurement of 'g' field intensity beyond the surface. The force per unit area at a distance 'd' measured from the center of mass is inversely proportional to the area $4\pi d^2$ where ($r_e < d < R$).

$$P_E = \left(\frac{M_e}{A_e}\right)(a_n), \text{ and } P_u = \frac{M_u}{A_u}g = (\sigma_u)g$$

For the pressure P_E to equal P_u at the space$-$mass interface, then:

$$g = \frac{P_E}{\sigma_u} = \left[\frac{M_e}{A_e}\right]\left[\frac{a_n}{\sigma_u}\right] = \left[\frac{M_e}{4\pi r^2}\right]\left[\frac{a_n}{\sigma_u}\right] = \left[\frac{c^2}{4\pi R\sigma_u}\right]\left(\frac{M_e}{r_E^2}\right)$$

identifying.......$\dfrac{c^2}{4\pi R\sigma_u}$ as big G...(26)

$$g = G\frac{M_e}{r^2} \text{...(27)}$$

For the earth $M_E = 5.98 \times 10^{24}$ kgm, $r_E = 6.37 \times 10^6$ meters. If $M_u/A_u = 1$ kgm/m^2, 'g' = 9.8 m/sec^2.

Calculation of the earth's 'g' field from (25) supports (but does not prove) the exactitude of the effective Hubble dynamic as **one kgm/meter^2**. Having established the value of big G per (26), the Newtonian formulation (27) naturally supersedes (25) for determining **g** fields of spherically symmetrical objects (the measured value of **G** being known with greater precision than **R**). What is revealed by (25), (26) and (27), is that which is missing from General Relativity, namely the root cause of gravity. Every 'g' field is due to the reactive counter force created by local matter in opposition to isotropic spatial expansion. Both the expansion acceleration a_n and the mass of the universe M_u are clandestinely encoded in the formulation of the acceleration source **G**. Each reactive field can then be individually defined by its own mass content M_e and the area A_e over which it is diluted. Local 'g' fields thus depend from **G** and **G** depends from Mach's Principle *a la* σ_U. The objections raised by Einstein are abrogated, the scalar density field acts instantly because Hubble mass transforms to a ubiquitous local property of the universe.

Per (23), the pressure of the void in a **q = -1** universe (which corresponds to $a_n = c^2/R$) having scalar density $\sigma_U = \rho_u R/3$ is:

$$(-P) = (c^2/R)(\rho_u R/3) = \rho_u c^2/3 \tag{28}$$

This is precisely what is required for negative pressure to cancel positive mass in Einstein's field equation.[11] The condition (28) is the only solution consistent with zero energy as the perpetual condition of the universe. Isotropic c^2/R acceleration is thus synonymous with (**q = -1**) de Sitter

---

[11] The empty de Sitter universe and the balanced energy universe (**-P = $\rho_u c^2/3$**) are governed by the same equation of state. Both expand exponentially. As revisited in equation (30) infra, σ_U must equal 1 **kgm/m^2** in a zero energy universe

expansion.[12] While pressure is uniform in a closed container *a la* Pascal's law, expanding negative pressure creates **g** field gradients in the vicinity of non expanding matter.

To recapitulate, we transformed the Hubble Mass to a surface density to create a 2-sphere. Some binding energy is lost in the process, so the 2-sphere is adjusted to be smaller by 5/6 which increases density sufficiently to restore binding energy lost in 3-sphere to 2-sphere transformation. The adjusted 2-sphere will have the same energy as the original 3-D Hubble sphere. The 2 sphere is then flattened to create a surface sufficiently large to approximate an infinite plane. The infinite plane can be mathematically located anywhere in the Hubble sphere because its influence is independent of the distance. Based upon the size and estimated mass of the Hubble sphere, the density of the infinite wall is approximately one **kgm/m²**.

How might Newton's reactionary Force arise from expansion? While motion through space is generally regarded as meaningless, the characterization does not apply to acceleration wrt space. Without an understanding of what space is, one would naturally assume that expanding space is likewise incomprehensible. But in fact, one property of space can be surmised.

Figure 1-B illustrates a matter shell constructed from disjoined pyramidal elements each having mass $\mathbf{m_j}$. In the first phase of our "thought experiment," the pyramids are deemed to accelerate outwardly along the radial at a rate $\mathbf{a_j}$ due to a positive interior pressure. The inwardly directed reactionary force $\mathbf{F = m_j a_j}$. The pyramids are then fixed to each other to prevent them from moving, and the experiment is altered by reducing both the interior and exterior pressure by the same amount until the interior pressure $\mathbf{P_i = 0}$. Consequently the exterior pressure $\mathbf{P_o < 0}$. The interior pressure now expands into the negative pressure of the void. What is the effect of the non expanding pyramidal surface upon the recessional flow of the spatial expansion field? Conservation of momentum requires reciprocity! Einstein reasoned "acceleration to be relative." The affect of non-expanding matter upon divergent spatial acceleration creates convergent counter forces throughout the universe. Newton called it "gravity." Einstein called the influence of inert mass upon static space "conditioning." But as is now known, space is not static. Mass conditions "dynamic space."

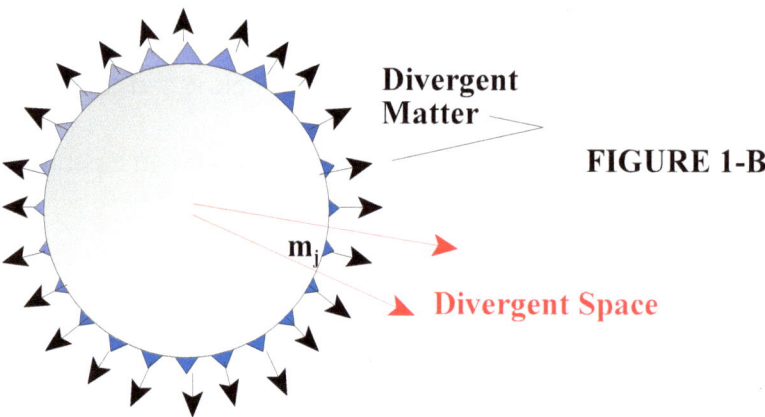

Divergent Matter

FIGURE 1-B

m_j

Divergent Space

---

[12]Transition from a more leisurely pace to accelerated expansion thus looms as a plausibility for a universe expanding from a hot dense genesis. If the present theory of accelerating expansion is ultimately validated, there is reason to consider the transition state as that which existed when decreasing energy density was overcome by negative cosmological pressure.

Figure 1-B is like the expanding universe. For energy to be zero, the pressure of space must be negative and of such volume and magnitude as to cancel positive energy. Negative pressure induces spatial expansion, but expansion of a negative pressure void creates positive energy. There being no real Hubble structure upon which negative pressure can operate, conceptualization reduces to the notion of momentum flow, i.e., Hubble recessional acceleration multiplied by the scalar density function $\sigma_u = \mathbf{kgm/m^2}$. Ironically, the universe is caught in a self creating construct. Expansion increases both negative and positive energy equally, the former by creating a larger volume to be occupied by the gravitational fields, the latter by enhancing inertia per (21). That inertia increases with size, so is it that Hubble density always appears to be hovering at the critical level ($\Omega = 1$). If the hand of God is at work, it is well hidden in the equations.

. A collateral implication of the above synopsis, sheds light on the issue as to why gravity depends from Hubble parameters rather than the size and mass of the greater unobservable universe beyond the Hubble sphere. If the ratio $\sigma_u = \mathbf{1\ kgm/m^2}$ is temporally invariant, the inertia of mass must increase per (21). The zero energy mandate determines the strength of the gravitational field; as a corollary the inertial property of positive matter must increase proportionately with the area of the surface that contains the volume.

****

The quest to derive **G** from first principles had its roots in the works of Alexander Friedmann (circa 1923) and later George Lemaitre (circa 1930). Both men, working independently, were able to formulate a model for an expanding universe in terms of the Hubble parameters **G**, **c**, **R** and $\mathbf{M_u}$. Prior to General Relativity, gravity was believed to emanate within matter. While Einstein disposed of the need for gravitational mass *per se*, he did so at the expense of creating a new physics principle based upon the proposition that inert matter conditions static space. But local 'g' forces depend from big **G** which is a dynamic ($\mathbf{m^3/sec^2}$ **per kgm**) herein matched to the expansion of space. General Relativity is silent as to the cause of gravity, and consequently makes predictions only after the measured value of the acceleration parameter **G** is introduced into the equations. Once **G** is known gravitational phenomena can be explained in terms thereof. No further postulation required. From Friedmann's equation and (17):

$$\rho_u = -\,\mathbf{q}\,\frac{3\mathbf{H}^2}{4\pi\mathbf{G}},$$

$$\text{then } \frac{3\mathbf{H}^2}{4\pi\mathbf{G}} \times \frac{\mathbf{R}}{3} = \sigma_u \tag{29}$$

$$\text{Therefore } \mathbf{G} = \frac{\mathbf{c}^2}{4\pi\mathbf{R}\sigma_u}$$

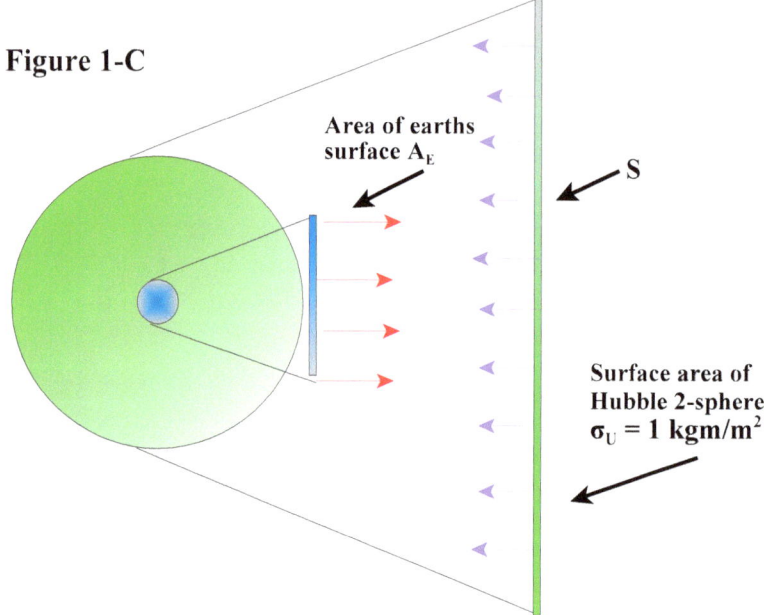

Figure 1-C

Area of earths surface A_E

S

Surface area of Hubble 2-sphere $\sigma_U = 1$ kgm/m^2

Figure 1- C A graphical 2-sphere tutorial deriving the 'g' field of a uniform density sphere, again taking the earth as exemplary. Total energy **M_u** (gravitational plus bare energy) contained within the Hubble sphere (Green) transforms to the surface area 'S' [flat plane surface density $\sigma_U = M_u/A_u$] where **A_u** represents an area which accounts for the energy lost in transformation.[13] For a zero energy universe, negative pressure **(-P)** must equal to $(\rho_u c^2/3) = [\sigma_U \times a_n]$ where **a_n** is the radial acceleration rate of recessional space **c^2/R**. Consequently, substituting **$Mu/[(4/3)\pi R^3]$** for **ρ_u** gives:

$$\sigma_U = (\rho_u c^2/3)/(c^2/R) = M_u/4\pi R^2 \tag{30}$$

Since the lines of action for the global force are perpendicular to '**S**,' Newton's 2nd law can be expressed in terms of densities. Specifically, the intensity of the global field (force per unit area) acting upon a uniform spherical mass (blue) is the global acceleration field divided by the density σ_U. The reactionary force equals the primary intensity multiplied by the mass of **E**. But since this force is spread uniformly over the surface of **E**, the force per unit area at the earths surface is:

$$\mathbf{g} = \frac{\mathbf{a}_n}{\dfrac{\mathbf{M}_u}{\mathbf{A}_u}}\left[\frac{\mathbf{M}_e}{\mathbf{A}_e}\right] = \mathbf{a}_n\,\frac{\sigma_E}{\sigma_u} = \mathbf{9.8\ ntn/kgm} \tag{31}$$

---

[13]As is the case for most astronomical bodies such as the earth, the gravitational binding mass-energy is insignificant compared to the bare mass, ergo, no adjustment is required when transforming the mass from a 3-sphere to a 2-sphere of the same radius. By contrast, the bare mass density of a zero energy universe is equal to its gravitational energy. The Hubble equation of state conforms closely with that of a black hole.

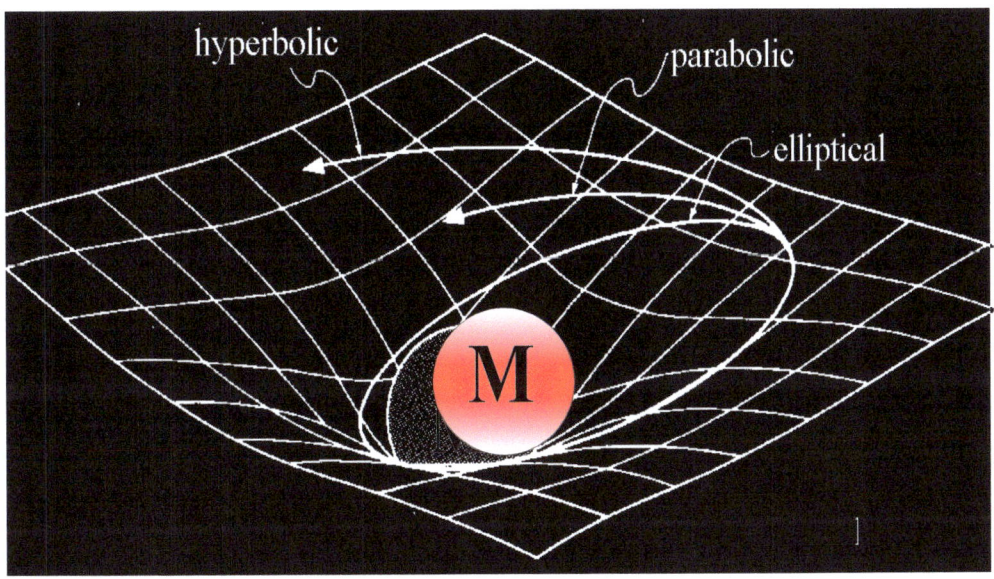

Figure 1-D: Illustrates the deformation of a metaphorical spacetime fabric when accelerated vertical as indicated by the two arrows A_n. Time runs slower and space is stretched..

NOTES////

///

The Variable Constants

Gravity-Mass Invariance

Mass, gravity, space, time and charge are intimately entwined. As one of two long range force fields, gravity will necessarily depend upon the content and action of the cosmos of which it is a part[1] Commencing with a 3-dimensional template built upon reasonable estimates of Hubble mass **(1.5 x 10^{53} kgm)** and size **(1.3 x 10^{26} meters)**, transformation to a 2-sphere shell model resulted in a reduced radius **(R$_2$ = 1.08 x 10^{26} meters)** and a consequent surface density **σ_u = one kgm/meter^2** By this artifice, Machian mass is insinuated into Hubble expansion to create isotropic momentum flow[2]

$$\sigma_u = (1.5 \times 10^{53} \text{ kgm})/4\pi(1.08 \times 10^{26} \text{ m})^2 \approx \text{ one kgm/meter}^2 \qquad (2\text{-}1)$$

Corollary, cosmic mass-energy can then be conveniently expressed per Part I (21) as:

$$M_u = 4\pi R^2 \text{ kgm/meters}^2 = 4\pi R^2 \sigma_u \qquad (2\text{-}2)$$

Figure 5 reconstructs the relationship between gravity and expansion, wherein the energy content of the Hubble sphere **M$_u$** is imagined as uniformly spread over the Hubble surface. If the empty interior is given a spatial dimension, the 2-sphere is reinvested as a 3-sphere. Per Feynman, "it costs nothing to create a mass at the center of a Hubble sphere (which is anywhere)." In a zero energy universe, positive **Mc^2** energy is balanced by negative gravitational potential. The interior can therefore be gradually populated with few chunks of matter from the shell while still retaining the notion of an inertial scalar density function **σ_u**. What is not in the shell is now in the interior volume fulfilling the job of creating '**g**' fields. Each mass is the center of its own Hubble sphere, it makes no difference where the shell is imaged. Transition from shell mechanics to 3-sphere homogeneity completes the ontogeny. However, as the interior density increases, the gravitational energy deficit also increases.

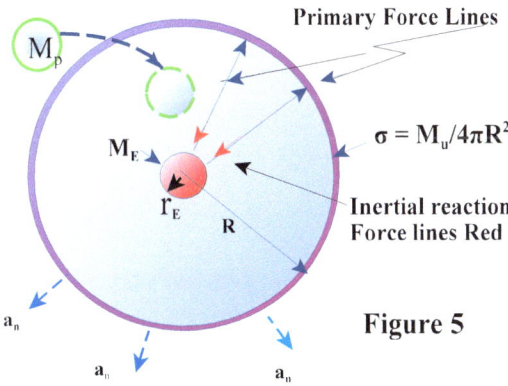

Figure 5

Primary Force Lines

$\sigma = M_u/4\pi R^2$

Inertial reaction Force lines Red

---

[1]When asked to summarize G.R. in one sentence Einstein Replied: *"Time, and space and gravitation have no separate existence from matter....physical objects are not in space, but these objects are spatially extended."*

[2]The expanding 2-sphere can be analogized to an inflating balloon where imaginary experiments can be made and results projected to the real world of three dimensional space. A true 2-sphere universe only has two dimensions. Our analogy of an empty interior 3-sphere borrows the 2-sphere formulation of energy as σ_u. The "Toy Model" in the context herein, is a set of parameters to be played-with to test alternatives

Expansion being yet undiscovered in 1916, Einstein inventively dismissed the lack of a force by postulating a new physics principle, namely: *"mass induced curvature of static space."*[3] While General Relativity correctly predicts the motion of masses along geodesics of curved space, it does so only because the empirical value of **G** is inserted to enable the functionality. Newton's **G** factor defines the trajectory, either directly as gravitational force created by the interaction of dynamic space and mass, or indirectly as mass created static curvature. Gravity as inertial reactance needs no interim curvature hypothesis.

In **Figure 5**, the black arrows illustrate isotropic spatial expansion c^2/R, the Hubble manifold being depicted as a stretching shell having perpetual surface density σ_u. The isotropic global acceleration field creates counter acceleration at the space-matter interface of any non-expanding inertial object [e.g., M_E as shown by red arrows]. In one of natures deceptive charades, reactionary '**g**' fields appear to originate internally.

Einstein proclaimed that masses act indirectly upon one another by making space unusual. To visualize how moving and non moving objects are influenced by the conditioning effect of mass upon space and time, it will be understood that all objects perpetually move through "spacetime" at velocity '**c**.' Those in a gravitational potential experience time passing slower, those moving through the gravitational field of a ponderable mass follow a straight line in curved spacetime called a geodesic. General Relativity makes accurate predictions once the measured value of **G** is inserted into the equations.[4]

When local '**g**' fields are viewed as counter actions, it is the **G** field that is bucked by inertial matter. Since '**g**' fields diminish inversely with area (same number of imaginary force lines cross each surface encompassing M_E), the sum over an encompassing surface at distance $r_1 > r_E$ is the same as any other encompassing surface at distance $r_2 > r_E$ Thus at a distance $r > r_E$:

$$F_r = [M_E(c^2/R)/\sigma_u][1/4\pi r^2] \tag{2-3}$$

As developed in **Part I**, field intensity for a mass M_E in an expanding three dimensional universe can be expressed in terms of the area over which the force is spread. Each mass reacts to the pressure $\rho_u c^2/3$ to create its own local '**g**' field proportionate thereto. Rewriting (2-3) as:

$$F = F_g = G[(M_E)/r^2] \tag{2-4}$$

where, as in the **Part I**, $c^2/4\pi R\sigma_U$ has been replaced by the symbol **G**. Without spatial expansion, there can be no force between σ_u and M_E. So also, without the reactionary mass M_u, the action of spatial acceleration a_n upon M_E creates no counter-force. The relationship between M_u and a_n is:

$$a_n = M_u G/R^2 \tag{2-5}$$

---

[3]To complete the theory, Einstein set up a static field equation, the left side he referred to as made of fine marble. It represented the scalar Riemannian manifold which described curved spacetime. The right side he dubbed a "house of straw." It premised matter as the cause. Only after discovering the elegance of the mathematical construct, did he artfully introduce a new role for mass as the provocateur of curvature. He was confident about the marble mansion, but never quite satisfied with the "house of straw.

[4]"God hath chosen the most foolish things of the world to confound the wise." (1st Corinthians I, vs 27).

Adverting again to **Figure 5**, mass beyond the surface of M_E does not add to the 'g' field of **E**. A second particle **P** introduced into **E's** Hubble arena, defines its own Hubble coordinate center. The reaction field of both M_p and M_E will be isotropic except to the extent M_E and M_p act upon one another. Global expansion thus converts Newton's 2nd law to volumetric expansion per unit mass resulting in the acceleration gradients that cause masses to be attracted to one another.

In **Part I** the scalar density function σ_u was used to derive the negative pressure field for the shell model. While volumetric spatial acceleration is the source, the essence of σ_u lies in the matter content of the Hubble sphere mathematically accorded inertial significance when transformed to an infinite flat plane. It will be understood, however that the density function need not be confined to a single plane. Multiple planes each having a density σ_u/n will be as good as one. This is the appeal of *"infinite plane mechanics."* For a number of infinite planes in every direction, the sum of the fields in any direction at every point will always be one **kgm/m²**.

While Hubble size and mass have diminutive physical significance individually, the scale **R** and mass M_u when taken together, determine the density of the infinite plane σ_u. In this sense, every Hubble sphere serves as a sampling artifice for the universe as a whole. Once **R** and M_u are determined empirically for the Hubble volume, the analytical value of **G** derived therefrom is applicable even to a flat space universe of infinite extent.

Volumetric growth of space $\dot{V}$ within the Hubble sphere and its derivative $\ddot{V}$ (volumetric acceleration) can be related to isotropic spatial flux $\dot{R}$ and its rate of change $\ddot{R}$. To find the internal production rate of space, the Hubble is enclosed by an imaginary Gaussian surface **S** of radius R_S. Accordingly, the following relations hold for **Figure 6**:

$$V = \frac{4}{3}\pi R^3 \dots\dots\dots\dots\dots\dots\dots\dots(2-6)$$

$$\dot{V} = (4\pi R^2)(\dot{R})$$

$$\ddot{V} = 8\pi R(\dot{R})^2 + 4\pi R^2(\ddot{R})\dots\dots\dots\dots(2-7)$$

Figure 6

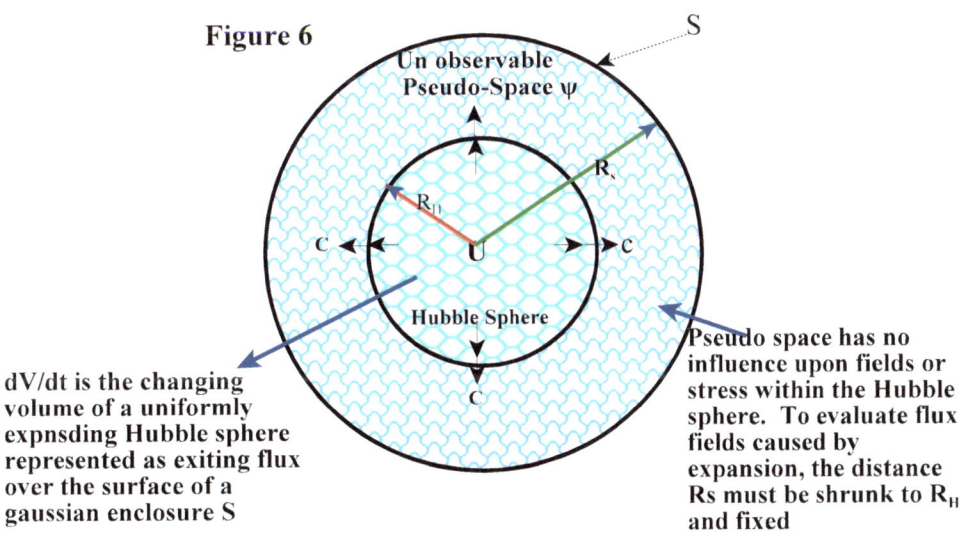

dV/dt is the changing volume of a uniformly expnsding Hubble sphere represented as exiting flux over the surface of a gaussian enclosure S

Pseudo space has no influence upon fields or stress within the Hubble sphere. To evaluate flux fields caused by expansion, the distance Rs must be shrunk to R_H and fixed

Encompassing the Hubble sphere with a Gaussian surround 'S' concurrent with the Hubble sphere at the instant of measurement is an adaptation of a volume to surface transformation first elaborated by the 18th century mathematician, Carl Friedrich Gauss.[5] For purposes of determining volumetric acceleration, the Hubble universe is considered devoid of mass, composed only of infinitesimal volumes, each expanding uniformly in three dimensions. In this expose´ volumetric expansion of space is treated as a functional operative at the smallest limit of existence (The vector divergence field is expressed mathematically as the fractional change in volume per unit area as the area approaches zero). Fractional change in volume per unit area can thus be regarded as a dynamic modulus, a measure of the intrinsic characteristic of space as expansion. Gauss's divergence theorem relates the integral over the volume of the surface that contains the divergences to the flux exiting across the surface that contains the volume. To apply the theorem to an expanding Hubble sphere, we sum over the exiting volume of recessional space and divide by the surface area of the Gaussian surround $4\pi R_S^2$. At coincidence, $R_H = R_S = R$, then from (2-7):

$$\frac{\ddot{V}}{\textbf{Area}} = \frac{8\pi R(\dot{R})^2 + 4\pi R^2(\ddot{R})^2}{4\pi R^2} = \frac{2\dot{R}^2}{R} + \ddot{R} \tag{2-8}$$

When (2-8) is expressed in terms of the deceleration parameter "q" then:[6]

$$\frac{\ddot{V}}{\textbf{Area}} = \frac{(\dot{R})^2}{R}[2 - q], \tag{2-9}$$

$$\textbf{where...} q = -\frac{\ddot{R}R}{\dot{R}^2}$$

In de Sitter's universe, $q = -1$, and therefore:

$$\frac{\ddot{V}}{\textbf{Area}} = A_H = \frac{3c^2}{R} = 3H^2R \tag{2-10}$$

---

[5] **Johann Carl Friedrich Gauss** (1777 — 1855) sometimes referred to as the *Princeps mathematicorum* (The Prince of Mathematicians)..

[6] After the discovery of the velocity-distance law **v = Hr** circa 1928, and throughout most of the 20th century, expansion was assumed to be slowing due to gravity. To express the rate of change in terms of velocity and distance, a **q** factor was concocted with a minus sign and given the name "deceleration parameter." If the rate of expansion is increasing with time, then **dv/dt** is positive and equal to **H(dr/dt)** or what is the same, **H²r**. At the Hubble distance **r = R** and **v = c**, so recessional flux exiting the Hubble sphere at any point is normal thereto. The Hubble surface is the transluminal locus of the **q = -1** universe where velocity is **c** and acceleration is (**c²/R**).

Equation (2-10) expresses volumetric acceleration per unit area of an exponentially expanding Hubble sphere in terms of the Hubble constant. But (2-10) is also Einstein's prescription for a static universe. More specifically, to balance gravitational force F_G tending to collapse the universe:.[7]

$$F_G = GM_u/R^2 = 4\pi G\rho_u R/3. \qquad (2\text{-}11)$$

Einstein's counter force Λ, originally intended to cancel gravity on the global scale, can now be justified as the essence of gravity. In the 100 years of experimental findings that followed its introduction in the final draft of his 1916 edition of the General theory, the Cosmological Constant has finally found respect, not as an artificially tuned opposition to balance gravity, but as its root cause. From the Friedmann-Lemaitre equations, this can be expressed as:

$$\Lambda R/3 = -4\pi G\rho_u R/3 = -H^2 R$$

And therefore: $$\Lambda = 3H^2 \qquad (2\text{-}12)$$

Exponential spatial growth is locked to 'c' at distance R where recessional space becomes transluminal, that is, where $c = HR$ and $a_n = c^2/R$. How is it, that Λ, can at once fix Einstein's universe as static while simultaneously sourcing exponential expansion? Metamorphosis from static to dynamic follows from the realization of ΛR as spatial expansion. For the mathematical model (General Relativity) to be static, the physiology must be functionally dynamic. When Einstein's Λ is embodied as spatial expansion, G emerges as $\Lambda/4\pi\rho_u$.

Expanding space is the implementation of Einstein's prescription for a balanced universe. As originally envisioned, it was an independent operative instituted to counter gravity. In reality the opposite is true. G is the manifest of expansion—and expansion is perhaps the natural consequence of negative pressure. Recognition of ΛR as the root cause of G renders the universe comprehensible.

---

[7]The same relationship follows if the M_E is placed immediately beyond the Hubble surface. M_E can now be considered a point mass separated from σ by distance R per (6). The velocity-distance law specifies the acceleration, i.e., if the rate of spatial expansion is increasing in proportion to the amount of space in existence, then since $v = Hr$, the acceleration is:

$$dv/dt = H(dr/dt) = Hv = H^2 r.$$

At the Hubble sphere, $r = R$, so the acceleration 'a' is $= H^2/R$. Equating this as the acceleration produced by the gravity field of the universe per(6) then:

$$M_u G/R^2 = H^2/R$$

Substituting the cosmic density-volume product $\rho_u V$ of the Hubble universe for M_u, there results:

$$G = 3H^2/4\pi\rho_u$$

This value is listed in the "Electronics World" table of constants. It obviously cannot be a constant because of the factor ρ_u in the denominator. Nor does it provide a physical model for G that can be used to rationalize the implied increase in G as the universe expands. [density is normally considered to diminish as $(1/R^3)$]. The derivation is not without merit however. When properly modified by a mass accretion algorithm, the above leads to the same value for G as the expanding two sphere model shown in **Figure 1**.

Einstein's inclusion of Λ in the 1916-1917 edition of the General Theory, and its resilience over the years, proves to be peculiarly fortunate. The recognition of Λ as the source of **G** resolves several enigmas, including the puzzling question of why density appears to be miraculously balanced between run-away expansion and gravitational collapse.

Any point can define a reference system for isotropic spatial expansion. The most convenient candidate for a coordinate center that would exemplify Richard Feynman's denotation of gravity as a *"pseudo force,"* is the geo-center of a uniform spherical mass. Most of what is considered as matter is actually empty space. Internal space expands, the centers of the atoms stay put because of internal binding forces. Einstein's Λ field is the isotropic flux that exits over the surface of the mass as illustrated in **Figure 1-B** of **Part I**. The inertial reaction of matter in reply masquerades as locally originated by gravitational mass, but as Einstein correctly observed, there is no gravitational mass *per se*. Inert matter in repose cannot self-create a force in static space. Inertial matter, can however, resist isotopic expansion, and in so doing, create **F = ma** counter acceleration intensities.

General Relativity outputs correct results, and with $\Lambda \mathbf{R}/3$ understood as spatial expansion, Einstein's theory of gravity is *complèter*. Gravitational and inertial mass are equivalent because both are inertial reactions. Λ acts upon matter to create isotropic inertial reactions, the intensity of the pseudo force field diminishes inversely with the square of the distance, while at the same time the area of the field increases with the square of the distance. Total reactionary flux is constant irrespective of distance (Intensity diminishes as $\mathbf{1/d^2}$, area increase as $\mathbf{d^2}$), consequently reactionary pressure acting upon σ_U is the same as that exerted by $\sigma_U(\mathbf{A_n})$ acting upon the surface of $\mathbf{M_e}$. As per [**Part I** (31)], if **E** is taken as the earth, then $\mathbf{M_E}$ denotes the earth's mass, and $\mathbf{r_E}$ its radius. To create the σ_E density shell, $\mathbf{M_E}$ is imagined uniformly distributed over the earth's surface to create a density $\sigma_E = \mathbf{M_E}/4\pi(\mathbf{r_E}^2)$. **Figure 7** depicts the Hubble surface density as a green circle $\sigma_u = \mathbf{kgm/m^2}$ and earth's surface density as a brown circle. As previously [**Part I** (20)], a cosmological acceleration factor $\mathbf{A_n}$ results in a local "**g**" force specified by the density ratio σ_E/σ_u where σ_u is one $\mathbf{kgm/m^2}$ and for de Sitter expansion, $\mathbf{A_n} = \mathbf{a_n} = \mathbf{c^2/R}$.

$$\frac{\mathbf{g}}{\mathbf{A_n}} = \frac{\sigma_e}{\sigma_u} = \frac{\dfrac{\mathbf{M_e}}{4\pi(\mathbf{r_e})^2}}{\dfrac{\mathbf{M_u}}{4\pi(\mathbf{r_e})^2}} = \frac{\dfrac{\mathbf{M_e}}{4\pi(\mathbf{r_e})^2}}{\dfrac{\mathbf{kgm}}{\mathbf{meters}^2}} \qquad (2\text{-}13)$$

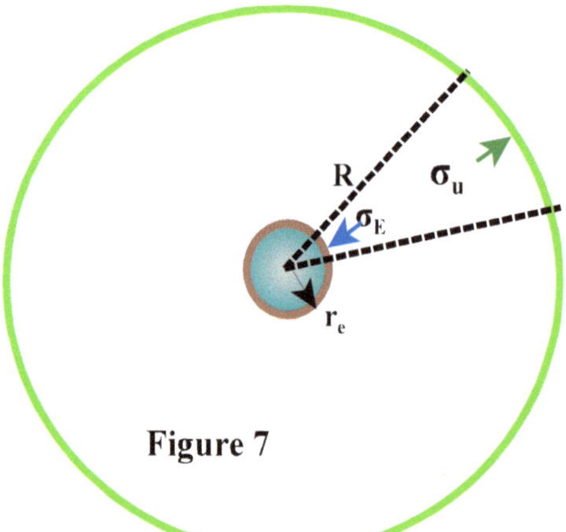

Figure 7

In the projection shown in **Figure 8**, the Hubble shell and Earth's surface are depicted as flat areas where the ratio $g/A_n = \sigma_E/\sigma_u$ as previously derived in **Part I**. Since $\sigma_E \gg \sigma_u$, 'g' is proportionately than A_n.[8] The ratio of surface densities (σ_E/σ_u) is equal to the ratio (g/A_n) of the acceleration fields.

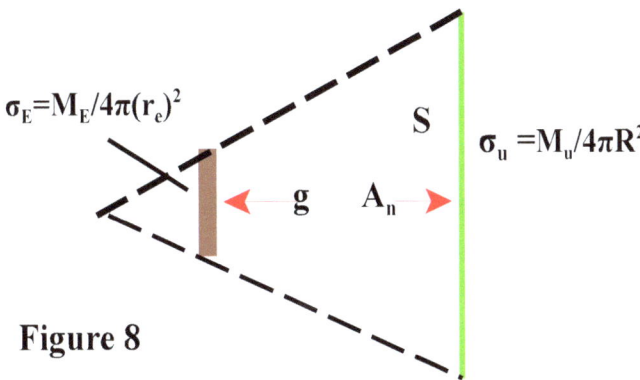

$$\sigma_E = M_E/4\pi(r_e)^2 \qquad S \qquad \sigma_u = M_u/4\pi R^2$$
$$g \qquad A_n$$

Figure 8

To address the central subject of **Part II**, namely the variability of **G**, attention is again directed to **Part I**, specifically the **R** term in the denominator (15) and (26), that is:

$$G = c^2/4\pi R\sigma_U \qquad\qquad (2\text{-}14)$$

The initial question posed by the **R** factor in the denominator of (1-14) is that of whether the Hubble radius is currently increasing. If not, any critique based upon the dependence of **G** upon **1/R** is unwarranted *ab initio*. Spatial flow at radius **R** is diverging at c^2/R.[9] If the de Sitter model of the universe is correct for the 'NOW" state of the universe, photons emitted in the direction of the earth at distance **R** from an earth observer will travel against the Hubble recessional field, presenting interesting questions regarding the energy lost and how it relates to the observed red shift.

Energy considerations also arise in connection with (2-14) if **R** is increasing The invariance of planetary Lunar orbits are generally regarded as placing sever constraints upon variable **G** theories. Orbital stability, however, depends upon the constancy of the MG product (where **M** is the mass of the larger object) which together with **G**, determine orbital velocity and radius. There are no test for measuring 'G' as a separate factor. If comic mass is somehow fixed at an early era as proclaimed by the majority of the physics community, the expansion dynamics must be finely tuned to explain the miraculous balance between density and kinetics. Changes in **G** alone cannot be detected because all such experiments require a mass upon which the gravitational field acts to produce the force. The dependence of **G** upon **1/R** does not invalid the herein ontogenesis.

---

[8]The present value of the Hubble constant H_o is usually expressed as recessional rate in (**km/sec**) per unit of distance measured in mega parsecs (**mpc**). One **mpc** = **3.09 x 10^{19}km**. For H_o = **70**, (or **2.3 x 10^{-18}/sec**) the Hubble time $1/H_o$ is = **3.09/70** or **4.4 x 10^{17} sec**. One year equals **3.16 x10^7 sec**, so the Hubble age is $\approx$**14 Gy**. The measured value of G (**6.67 x 10^{-11} m^3/sec^2 per kgm**) corresponds to H_o in the range of **70. km/sec/mpc**

[9]If the initial velocity of a photon created at the Hubble radius is 'c' then the energy lost in traversing the distance to the earth is hf.

The dependence of **G** upon **1/R** and the approximately linear dependence of Hubble scale upon time, is consistent with Dirac's large number hypothesis. In Dirac's theory, **ΔR/R** must greatly exceed the small variation of **G** allowed by orbital uncertainty.[10] The constancy of the **MG** product requires that **G** and M both be constant, or **M** must increase as **G** palliates. Already then, any theory of variable **G** also requires a corollary, in the form of gradually acquired inertia. The acceptance of a change in one demands a compensatory change in the other. Two variables are involved, both of which are deemed constant by the keepers of the Standard Model. That the inertia of individual masses increases with time also requires that total cosmic mass energy increase in proportion to the area of the Hubble manifold per (21).[11] That $\sigma_U = 1$, the dictates of zero energy and the inverse variability of gravity and mass become operationally necessary for internal consistency in an evolving structure. A universe that acquires mass in a short instant cannot be a player in the zero energy schema.

Attempts to measure **G** separately are masked by the constancy of the **MG** product.[12] As predicted in **Part I**, an increase in surface area is accompanied by an increase in mass and consequently, a decrease in the value of **G,** but not a decrease in the magnitude of the negative energy defined by the gravitational field. During expansion, the volume of the negative pressure field increases as **R³** and **G** decreases by **1/R** so negative energy contained within the expanding dimensions of the Hubble volume increases by **R²**. Negative energy increases at the same pace as Hubble area **4πR²**, so cosmic mass **M_u** can be written in conformity with (21) as **4πR²(σ_u)**.[13, 14, 15]

---

[10]Paul Dirac proposed that the near equality of certain large numbers could be easily explained if **G** varied as **1/R**. But for **G ∝ (1/R)**, the inertial reactance of matter must increase proportionately.

[11]While Energy is conserved, mass converted to other forms need not exhibit inertia, and when in the form of mass, moving at relativistic velocity, greater inertia is observed. Mass is a state of energy, not a conserved quantity. Theories is that require a change in one parameter lead to a change in others.

[12]For an orbital moon, **v** and **r** depend upon the product of the planet's mass **M_c** multiplied by **G**. The orbital parameters **v** and **r** are determined by equating **GM_cM*/R²** to centripetal force, **M*v²/r** from which:

$$GM_c = rc^2 \qquad\qquad (2\text{-}14)$$

[13]As an aside, gradually acquired inertia resolves a cosmological quandary, namely the high degree of tuning required for critical density. **G** has units of volumetric acceleration per unit mass **[m³/sec² per kgm]**. To what might these units apply other than expanding space? And if applied to expanding space, is it not conceivable that the rate of volumetric growth would change as the size of the universe changes? The stability of orbits is testament to the invariance of the **MG** product. If **G** palliates as **1/R**, the mass factor in the force equation must increase proportionately with **R**.

[14]In 1937 Paul Dirac published the Large Number Hypothesis (**LNH**). Reasoning that the near equality between the electro/gravitational force ratio and Hubble/subatomic size ratio must be more than a coincidence, Dirac suggested that these large numbers maintain the same proportions at all times. This can only be true if one of the so called constants of nature changed as the universe expands. This lead to Dirac's hypothesis that **G** varies as ∝ **(1/R)**.

[15]There is no law of conservation of mass. The inertial resistance of masses to acceleration increases for masses traveling at high velocities relative to the fame of measurement. Nor is their bases for the idea of an explosive mass creating genesis, although much effort has been directed to justifying such scenarios. Gradually acquired inertia is as it must be, it must grow to balance the negative energy of the expanding Hubble volume.

Hubble energy is balanced to zero so long as negative and positive energy increase conjointly (Positive matter created in the form of inertial enhancement equals the gravitational energy in the new volume occupied by the negative pressure field). In a zero energy universe, excess positive energy must be carried off by increased recessional flow, the excess energy itself being that required to increase the rate of spatial recessional flow exiting across the Gaussian surface $4\pi(R_s)^2$ depicted in **Figure 6**. Once positive matter density exceeds negative potential energy, exponential expansion becomes self sustaining. This may be an initial condition or a later condition. If later, the net zero model that fits exponential expansion must be modified accordingly for earlier periods.

Prior to Einstein's recognition of equivalence, inertial and gravitational mass were viewed as separate but enigmatically equal functionality(s). Herein, reactionary '**g**' fields supersede static distortion. With no separate existence from that which brings about their existence, understanding comes from an acceptance of the expansion field as the source of gravity.

Following Hubble's discovery of the *velocity-red shift law* and subsequently, Howard Robertson's formulation of the *"velocity-distance law,"* British Cosmologists, William McCrea and Edward Milne, expressed their doubts as to the viability of Einstein's presumptive mechanism. Gravitation as a warping of space, although a credible notion in and of itself, offered not a hint as to how it could come about. Both men worked to derive the same gravitational equations that had been previously extracted from the General Theory. Their results were based upon energy considerations alone, and formulated entirely from Newtonian dynamics. What is revealed is that the General Theory is in essence, an energy theory. For the non professional who wishes to understand gravity without delving into mathematical complexities, the equations are as follows:

$$\dot{R}^2 = \frac{8\pi G R^2 \rho_u}{3} + \frac{\Lambda R^2}{3} - \frac{kc^2}{R^2} \tag{2-15}$$

$$\ddot{R} = -\frac{4\pi G}{3}\left[\rho_u + \frac{3P_s}{c^2}\right]R + \frac{\Lambda R}{3} \tag{2-16}$$

where k/R^2 is the curvature. Equations (2-15 and (2.16) are identical to the equations first synthesized from the Theory of General Theory by considerable skill and labor. That Newtonian physics should lead to the same equations is nothing short of sublime. In both derivations, the density is understood to include all forms of mass energy in all possible forms (e.g., bare mass, Komar mass, gravittional mass, radiation, thermal, kinetic energy, relativistic etc). From (2-16), it is now obvious why negative pressure having a magnitude

$$(-P = \rho_u c^2/3) \tag{2-17}$$

greatly simplifies the matter dominated present state of the universe. Energy is net zero, and (2.16) reduces to:

$$\ddot{R} = \frac{\Lambda R}{3} \tag{2-18}$$

Which has the same solution as de Sitter's empty universe, i.e.,

$$\frac{3\dot{R}^2}{R^2} = \Lambda \qquad (2\text{-}19)$$

And therefore

$$\frac{\dot{R}}{R} = \left[\frac{\Lambda}{3}\right]^{1/2} \qquad (2\text{-}20)$$

Whence:

$$R = e^{\left[\frac{\Lambda}{3}\right]^{1/2} t} \qquad (2\text{-}21)$$

From (2-12), Einstein's Λ substituted into (2.15) can now be interpreted as synonymous with c^2/R acceleration, with G the emergent consequence. When $\Lambda = 3H^2$ is substituted into (2-21), then:

$$R = e^{Ht} \qquad (2.22)$$

As an aside, an alternative model of the universe obtained from (2-16) is:

$$P = -\rho_u c^2 \qquad (2.23)$$

then (2-15) and (2-16) become (24) and (25):

$$\Lambda = -8\pi G\rho_u - 3qH^2 \qquad (2.24)$$

$$k/R^2 = -H^2(q + 1) \qquad (2.25)$$

The state equation $P = -\rho_u c^2$ was seized upon Fred Hoyle and other detractors of Big Bang Theory as an underpinning for "*Steady State Theory*." Positive energy created by expansion of negative pressure volume maintains density constant. The universe has neither a beginning nor and end. This was McCrea's ingenious idea for creating matter.[16] In 1981, Alan Guth appropriated McCrea's recipe to develop his own theory of expansion which he called "*Inflation*. Like the $-P = \rho_u c^2$ universe, constant density expansion also has an exponential solution, but it does not meet the zero energy requirements of our ontogeny.

---

[16].McCrea's concept exists in one form or another in most creation theories. In contrast to Hoyle's "*Steady State Theory*," the doctrine of gradually acquired inertia foretold herein, requires no new particle production.

As previously developed, in Part I, for zero energy:

$$(\rho_u c^2/3) = (-P) = (c^2/R)\sigma_U \qquad (2.26)$$

And since

$$\rho_u = 3\sigma/R = M_u/(4/3\pi R^3)$$

Then

[as previously from (21)]

$$M_u = 4\pi R^2 \sigma \qquad (2.27)$$

All of which leads back to (7):

$$\frac{M_u G}{c^2 R} = 1 \qquad (2.28)$$

The ratio (2-28), once considered an inscrutable cosmological enigma, can now seen as an identity for any value of **R**:

$$\frac{M_u G}{c^2 R} = 1 \quad = \quad \frac{[(4\pi R^2)\sigma_U]}{(c^2 R)} \times \frac{c^2}{4\pi R \sigma u} \qquad (2.29)$$

Equation (2-29) holds for any value of **R**, consequently it applies to any era. This raises the issue of how radiation energy is balanced? Photons are massless, and consequently create no gravitational field energy to negate photon **hf** energy. Photon pair production from expanding negative pressure preserves the state of cosmic momentum as zero.[17] Consequently the energy of radiation in one direction will balance the energy of radiation in the opposite direction. However, both energies are positive - even though momentum is balanced, total KE ≠ 0. However, in a cosmic system where all energy is in the form of radiation, the size of the operative structure will be smaller than a matter dominated universe having the same number of individual energy elements. Ergo, as matter distills from the soup, the universe must expand rapidly to create negative pressure volume energy to balance newly created mass (In general, a greater volume is required to accommodate the negative pressure field ($-P = \rho_u c^2/3$).[18] In standard theory, Hubble expansion appears to have been better understood after the decoupling era. In this treatise, however, the objective is that of deriving the present **G** factor in terms of the present conditions of the universe. This requires M_u increase as R^2 whereas the inertia of individual particles must increase linearly with **R** to offset the greater negative potentials created by expanding distances.

---

[17]For example the union of an electron and positron create a pair of gamma ray photons. Ignoring the motion of the electron and positron prior to annihilation, the gamma ray photons will be oppositely directed.

[18]If the present mass of the universe were contained withing a volume of 10^{53} cubic meters to create a density of one **kgm/m³**, the negative pressure gravitational field will occupy a volume 10^{26} times greater.

Here we conclude our derivations of **G** with a simply adaption of Gauss's Theorem to the Hubble sphere. Thus if $\mathbf{M_u}$ is the effective mass and **R** is the effective radius, '**g**' flux exiting across the surface corresponds to the expansion acceleration factor $\mathbf{A_n = c^2/R}$ then:

$$\int_S \mathbf{g} \cdot \mathbf{n}(\mathbf{dA}) = -4\pi\mathbf{GM_u} \tag{2.29A}$$

Whence

$$\mathbf{G} = \left[\frac{\mathbf{c^2}}{\mathbf{R}}\right]\frac{4\pi\mathbf{R}}{(-4\pi)\mathbf{M_u}} = \frac{\mathbf{c^2 R}}{\mathbf{M}_u}$$

And for $\mathbf{M_u = 4\pi R^2(\sigma_U)}$

$$\mathbf{G = c^2/4\pi R\sigma_U}$$

SPECULATIONS

A. The Reality of Cosmological Acceleration

While evidence of *Accelerating Spatial Expansion* during the last 6 billion years is strong, the overall profile of Hubble growth is nearly linear over what is considered the life of the universe. This would appear to discredit exponential growth as an initial condition. However, as with all efforts to retro-model the past based upon the present, the assumptions regarding the growth of "*space*" and the passage of "*time*," as judged by the "*now*" state of the Hubble sphere, may be wrong. If temporal intervals are self referential, the universe may have existed forever while nonetheless portraying the countenance of a genesis. The continuity of the equations derived from the zero energy mandate demand perpetual expansion. If a zero energy symmetry can take effect as a universe, it must take effect as an exponential expansion, or not at all. To illustrate, if a spatial interval 'S' is considered to arise during a kinematic temporal interval "t" $= 1/H_r = r/c$ where 'a' is the cosmological acceleration c^2/r, 'r' is the Hubble scale at time 't' and τ is the current Hubble age measured by the current rate of aging, then:

$$\mathbf{S} = \int \mathbf{dS} = \int^\tau \mathbf{at(dt)} = \int^\tau \frac{\mathbf{c}^2}{\mathbf{r}}\mathbf{(t)dt} = \int^\tau \left[\frac{\mathbf{c}^2}{\mathbf{r}}\right]\left[\frac{\mathbf{r}}{\mathbf{c}}\right]\mathbf{dt} = \mathbf{c}\tau \qquad (2\text{-}30)$$

For kinematic time ($\mathbf{c/r}$) the expansion profile is linear in time.

The approximate linearity of the distance time graph calls into question the common assumption that recessing nebula are commoving wrt the spatial flux. From the perspective of an earth observer, Hubble recessional flow is isotropic. A galaxy at the Hubble limit recedes at velocity 'c' while urged by the accelerating rate of spatial flow to accelerate at c^2/R to maintain its presumed comoving state. However, taking the cosmic mass $\mathbf{M_u}$ is $4\pi R^2 \sigma_u$ then the retarding effect of Hubble mass acting upon a receding galaxy of mass $\mathbf{M}$ at the Hubble limit is:

$$\mathbf{F}_G = \frac{4\pi \mathbf{R}^2 \sigma_u [\mathbf{MG}]}{\mathbf{R}^2} = \frac{4\pi\sigma_u \mathbf{c}^2 \mathbf{M}}{4\pi \mathbf{R}\sigma_u} \qquad (2.31)$$

Equation (2.31) reduces to $\mathbf{F} = \mathbf{Ma}$. The gravitational force per unit mass $\mathbf{c}^2/\mathbf{R}$ is equal and opposite to the recessional acceleration field of space at the Hubble distance. Net force acting upon a mass at distance $\mathbf{R}$ is therefore zero. Since zero net force produces no acceleration, no dark energy is required by way of explanation. Massless spatial acceleration appears to be the present condition of the universe.[19]

---

[19]If the masses are not comoving and therefore not accelerating, how might the apparent diminution of the more distance supernova events be rationalized?

B. The Faint Supernova Studies

Spontaneous creation has been a recurrent theme throughout scientific history. But an abrupt beginning cannot include the entire mass of the universe in a single event. The sudden appearance of mass-energy out of nothing is contrary to the zero energy scenario, and discrepant with all that is known about evolutionary process. Nonetheless, the general sentient of the twentieth Century had the more distant galaxies receiving a greater initial boost and thereafter all galaxies were assumed to be slowed by gravity. The all at once matter myth requires expansion velocity to be fine tuned to avoid a quick crash or runaway expansion.

The 1998 supernova studies were based upon the proposition that SN bursts could be used as standard candles–the exclamation of identical energies, and therefore of equal brightness and duration. To the investigators's surprise, the intensity of the more distant events appeared fainter than what would be expected for a slowing universe. Either the universe was accelerating or something else was in play in the distant past.

The gravitational pressure needed to trigger a supernova was derived in 1932 by the Indian physicist, Subrahmanyan Chandrasekhar, for which he later received the Nobel prize.[20] The critical energy M_{limit} (approximately 1.4 solar masses) depends upon the factor $(hc/4\pi G)$. If **G** diminished inversely with **R**, the invariance of the **MG** product speaks directly to the question of whether supernova events were less energetic in the past. If that be so, the evidence for exponential expansion of matter vanishes, and so also does the search for dark matter.[21] The dimming of the intensity however can be rationalized by a larger **G** factor in the past. Consequently for the MG product to be constant, less inertia is requred to create the same weight force.[22] Since electron degeneracy pressure (**hc**) is constant, the inertial factor is less (Because **G** is greater in the past, less inertial matter is required to trigger a **1a** supernova event in the early universe). If intensity diminution is the result of less mass rather than greater distance, the theory of accelerating expansion is erroneous. In this thesis, exponential cosmological expansion is the auspicate of declining **G** and its corollary, the gradual acquisition of inertia.

---

[20] A white dwarf star is stabilized by two opposing forces: 1) electron degeneracy pressure created by nuclear fusion in the heart of the star (making lighter elements into heavier ones) pushing outwards from the core, and 2) gravity pulling inwards. When a white dwarf is locked in orbit with a companion star, it sucks off matter. Over time this increases gravitational pressure until it overcomes the electron degeneracy pressure. The mass in the core has a special significance called the Chandrasekhar Limit. When the core acquires a mass of approximately 1.4 solar masses, the electron degeneracy pressure is overcome by the pressure of gravity acting upon the core.

[21] As a side note, efforts to explain the present value of **G** in terms of $q = \frac{1}{2}$ led to much frustration for the author. The discovery of Cosmological Acceleration provided a good fit to the empirical value of **G** based upon standard model consensus $H_o = 71$. The perception of uniformly expanding **3-D** space as 'time' can be appreciated as a consequence of a changing **4th** dimensionality.

[22] Because the **MG** product is constant, the weight of the mass required to overcome the degeneracy pressure is the same at all eras. When the weight overcomes the electron degeneracy pressure, the white dwarf star collapses with a violent luminous display. Since the electron degeneracy pressure does not change with time, the **MG** pressure required to trigger a supernova event will also be invariant irrespective of the individual contributions of **G** and **M**. A robust **G** during an earlier era translates to smaller **M**, and consequently less energetic events. For the present value of **G**, Chandrasekhar's equation predicts 1.4 solar mass as the critical value.

C. The Faint Sun Syndrom

Geophysical and climatological data show the earths temperature during the past four billion years has not changed appreciably. However, numerical models based upon the Sun's interior indicate the Solar output would have been approximately 25% less than its present value. Various theories have arisen to justify the warm conditions that prevailed for the young earth. Of significance for this treatise, is the variable **G** theory.

The Sun's luminosity $L_\odot$ is highly sensitive to **G** and **M**, being roughly proportional to G^7M^5. The invariance of the **MG** product thus provides the mechanism and explanation for a temperate beginning. A 25% reduction in solar output based upon the Sun's condition and status as a main sequence star would be roughly balanced by a robust **G** and a smaller inertial mass **M**. In a recent publication, the variable **G** theory was studied and compared to alternatives founded upon atmospheric changes, most notably suppositions based upon unsupported levels of green house gases in the atmosphere of the young earth.[23] The authors avoided consideration of changes of size proposed by Dirac' in his Large Number hypothesis ($G \propto t^{-1}$). Consequently, they were able to explain their conclusions in terms of small changes. However, as developed herein, decreasing **G** is accompanied by increasing inertial mass, the effect upon luminosity $L_\odot$ due to changes in **G** is reduced from 10^7 to 10^2 when the 10^5 effect of increased mass is factored into the equation.

---

[23]Can A Variable Gravitational Constant Resolve The Faint Young Sun Paradox?; International Journal of Modern Physics D. Varun Sahini, Yuri Shtanov, Nov 2014.

D. Relativistic Conformance.

In what mistaken appears to be a mechanistic disconnect between time dilation in Special and General Relativity, in reality both depend upon the energy state of the clocks used to make comparisons. Although rarely expressed as such, real differences always involve energy differences. For Special Relativity, the apparent rate at which measuring clocks accumulate '*time*' depends upon relative velocity. In the General Theory, a clock in the gravitational potential of a mass **M** accumulates time slower than the same clock far removed. However, the energy required for removal from the potential well created by **M** is equal to the escape velocity:

$$\Delta t^* = \Delta t (1 - 2GM/rc^2)^{\frac{1}{2}} \tag{2.32}$$

where **2GM/rc²** is the escape velocity needed overcome the gravitational field of **M**, or viewed alternatively, the velocity acquired by an object falling from ∞ to the surface of a uniform spherical mass **M** of radius **r**. That this factor has a familiar complexion, suggests implications applied to derive **G** in terms of $\mathbf{M_u}$ and **R**, that is:

$$\Delta t^* = \Delta t \left[1 - 2 \frac{4\pi R^2 \sigma_u c^2}{4\pi R \sigma_u (Rc^2)} \right]^{1/2} = \Delta t [-1]^{1/2} = i\Delta t \tag{2.33}$$

Hence (2.32) reduces to one unit defined along the "**i**" axis ⊥ to three dimensional space.

What is revealed is that real time dilation depends upon real energy differences – clocks placed in oppositely moving spaceships $\mathbf{J_1}$ and $\mathbf{J_2}$ log identical 'times' with respect to an intermediate inertial frame $\mathbf{E_3}$ from which each is given an equal but oppositely directed acceleration boost (red). Observers in the $\mathbf{J_1}$ and $\mathbf{J_2}$ spaceships, however, will falsely measure the clock in the other spaceship to be running slow.

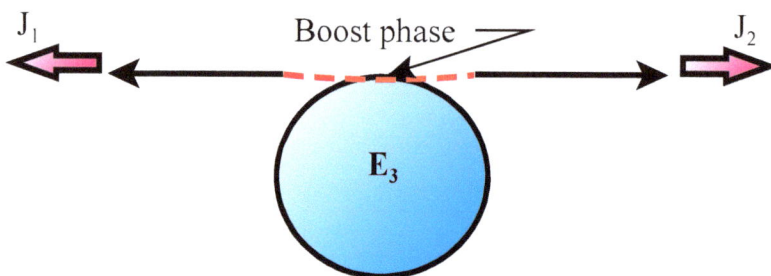

F. Dark Energy Not Required!

Taking Newton's 2nd law as prescriptive for zero energy, then $\int \mathbf{F} \cdot \mathbf{ds} = 0$ when $\mathbf{F} = 0,$

$$\mathbf{F} = d/dt(Mv) = M(dv/dt) + v(dM/dt) = 0$$

wherefore for an expanding constant density ρ_u

$$M_u = \rho_u(4/3)\pi R^3$$

then
$$dM/dt = \rho_u(4\pi R^2)(dR/dt)$$

From which
$$\mathbf{F} = (\rho_u)(4/3)\pi R^3)(dv/dt) + (v)\rho_u(4\pi R^2)(dR/dt) = 0$$

At the hubble limit, $\mathbf{dR/dt} = \mathbf{c}$. For zero net force:

$$\frac{dv}{dt} = a_n = -\frac{v^2(4\pi R^2)\rho_u}{\frac{4}{3}(\pi R^3)\rho_u} = -\frac{3c^2}{R}$$

Thus for the constant density universe, cosmological acceleration equals $3c^2/R$ so:

$$-P = (3c^2/R)\sigma_u = \rho_u c^2$$

Whence:

$$\rho_u = (3\sigma_u/R)$$

For negative pressure $-P = \rho_u c^2/3 = 3\sigma_u/R = a_n\sigma_u$, then cosmological acceleration is:

$$a_n = \rho_u c^2/3 = 3\sigma_u/R = c^2/R$$

$$\frac{\mathbf{F}}{\mathbf{M}_u} = \frac{M_u\ddot{R} + \dot{R}^2\rho_u[4\pi R^2]\dot{R} + M_u\dot{R}[\frac{d\rho_u}{dt}]}{M_u}$$

For $q = -1$

$$\dot{R}^2 = \ddot{R}R, \text{ and since } \rho_u[4\pi R^2]/M_u = 3/R, \text{ then}$$

$$\dot{\rho}_u = 2\frac{\dot{R}^2}{R^2} = 2H^2 \text{ and therefore } \rho_u \propto \frac{1}{R}$$

Conclusions

The evolution of the universe will be guided by exponential expansion when negative pressure $(-P = \rho_u c^2/3)$. Whether this condition has always been a factor in the evolution of the universe, or if it arose as a later state, is not known with certainty. Based upon **1a** super nova studies, a period in the range of **7** billion years past, would conform to what has been estimated as the onset of exponential expansion. But '**0**' energy must be satisfied at all eras; and a different model will be required during any period not governed by exponential growth.

The object herein is to correlate the *'present'* value of **G** with the *'now'* rate of "spatial expansion." From a dimensional perspective, Newton's "Law of Gravitation" follows from his "2nd law of motion." Spatial acceleration multiplied by local mass defines local '**g**' fields. The speculations of Richard Feynman have been carried to a testable conclusion. The value of **G** is predicted by the expansion rate of space

Bare mass-inertia increases linearly as Hubble scale **R** increases. Gravitational binding energy increases proportionately with bare mass. Total mass-energy therefore increases as R^2.

What affect does all this have upon General Relativity? Actually very little, save for the fact static spatial distortion is replaced by dynamic reaction of expanding momentum space. Although originally envisioned by Max Born in terms of a gateway to unification with quantum physics "*Born reciprocity*" was developed as a direct result of the mathematical similarity between the dynamic effect of mass upon momentum space and the prophesied distortions of mass upon static space.

General Relativity was originally built upon Riemann geometry. Pieced together surfaces from patches of the complex plane, they define a topological manifold of great utility. But given the evidence for expansion, nothing further is required to explain gravitational action? A global acceleration field distorted by local matter is a natural and necessary result of the symmetry encoded in the principle of action-reaction. General Relativity did not attempt to derive or explain **G**. Despite its successful predictions, it remained unfinished. The predication **G** of upon **Λ**, completes Einstein's Ontogeny.

We begin our expose' with a quote from Richard Feynman. It is fitting we close it likewise:

"No machinery has ever been invented that explains gravity without also predicting some other phenomena that does not exist."

****

Readers are invited to email comments to: © B. D. Jimerson, 2015.
cosmodynamics@yahoo.com

HARD TO BELIEVE PINCOCK, BUT FROM THE MATHEMATICAL
PERSPECTIVE, THE UNIVERSE BOILS DOWN TO NOTHING
GOING GOD KNOWS WHERE IN EVERY DIRECTION

Notes

PART II

The Origin of the Electric Field

Prelude to Charge

Creation theories are built upon Space, Time and Mass. The electric field receives neither explanation nor mention in Big Bang scenarios or steady state theories. Indeed, it is overlooked in biblical genesis as well as other folk tales based upon revelation. What then, is electrical charge? The precepts of physics are replete with evidence of an underlying communion between the two long range force fields. Despite centuries of thought, neither has been derived, one from the other, nor has a common root been found from which they might be synthesized.

The challenge is a seductive one, luring many of the great minds to its intrigue. More than a half century before Einstein took up his theoretical quest, Michael Faraday sought enlightenment in the laboratory. In 1849 he scribbled these words in his notebook:[1]

> *"Gravity: Surely this force must be capable of an experimental relation to electricity, magnetism, and the other forces, so as to bind it up with them, in reciprocal action and equivalent effect."*

After many unsuccessful experiments he concluded:

> *"Here end my trials for the present. The results are negative. They do not shake my strong feeling of the existence of a relationship between gravity and electricity, though they give me no proof that such a relationship exists."*

The advantage of today's inquisitor lies in the knowledge of space as expanding. A key factor in the derivation of **G** is the physiol adaptation of isotropic global acceleration into an operative. To extend this to the interaction between electrical charges requires the a new perspective as to what can be claimed as a reactive force. The electric force between electrons and positrons exceeds their gravitational attraction by 42 magnitudes [10^{42} greater than that created by the **G** field acting upon the mass of the electron.[2] The **G** field is weak because the present value of **R** is large. For an electron sized universe, the **G** is comparable to the electric field. As Newton professed two plus centuries past, forces are accelerations. To find large accelerations in a big universe requires a new model of the small universe.

That the **G** field depends upon Hubble scale, should cause no surprise. Since most Hubble factors cannot be directly measured, estimates are made from measurements of brightness and red shift. The derivation of the electric charge from the electron size and mass also appears to be an approximate proposition, but there is a critical difference in that subatomic properties are quantized. Thus, while the rules of quantum theory define the interaction between charges, the laws of classical physics control the formulation. In what follows, the electron will be considered the defining structure for one unit of charge. While other entities appear to display identical electrical properties, distinctions will be reserved to simplify the analytics.

---

[1]Encyclopedia Britannica, 1971, pp. 670, 673

[2]Herein, the electron mass is denoted as m_o defined by the amount of energy required to be deposited on a spherical surface of radius $r_o = 1.41 \times 10^{-15}$ meters to create a charge equal to one electron 'q'

Like many other subatomic particles, electrons exhibit angular momentum $L = h/4\pi$ along any axis of measurement. Viewed from the classical perspective of angular momentum as rotating mass, consternation follows:

"....we described how in quantum mechanics the angular momentum of a thing does not have an arbitrary direction, but its component along a given axis can take on only certain equally spaced, discrete values. It is a shocking and peculiar thing ...There isn't any descriptive way of making it intelligible that isn't so subtle and advanced in its own form that it is more complicated than the thing you were trying to explain....Understanding these matters comes very slowly, if at all...the most shocking and disturbing thing about quantum mechanics is that if you take the angular momentum along any particular axis you find that it is always an integer or half integer times h/2π."

Richard Feynman, Lectures On Physics

Dimensionally, a particles moment of momentum is defined by the cross product, specifically in classical physics, angular momuntum is symbolized by $L = p \times r$ where p is the linear momentum vector (mv) and 'r' is the radial vector. Expansion acts upon all forms of bound energy to create reactive counter action. If local gravity fields virtualize as spatial divergence created by non-expanding matter, a fixed unit of angular momentum emulates as counter vorticity. Denoting electron mass as m_o, the energy in the circulatory counter field is $m_o c^2$ and the classical angular momentum $L = m_o c r_o$ where $r_o = 1.41$ **fermi** [(½) the classical radius r_e], the result of equating the electron mass m_o to the energy required to form a spherical surface of charge $q = 1$ electron. Just as spatial expansion acts upon bound matter to create divergent counter reactive 'g' fields over limitless volume, expansion acts upon local angular momentums to create vortical counter fields. Unlike classical vortices, the electric counter field is three dimensional and velocity at any distance from the eye equals 'c' And while the total energy of the counter rotational field equals $m_o c^2$, total spatial angular momentum is $h/4\pi$ and not $m_o c r_o$.[3] Adapting Gauss's law for gravitational flux(equation 2.29A) to an angular momentum entity, with 'j' substituted for 'g' as the exiting momentum flux and (angular momentum x spatial expansion) substituted for Gm, then:

$$\int j \cdot n \, dA = -4\pi (m_o r_o c) \frac{dR}{dt}$$

$$j = \frac{-4\pi m_o r_o \frac{dR}{dt}}{4\pi R^2} = -\frac{m_o r_o c^2}{R^2} \text{ and for any other } dis \tan ce \text{ 'd'}.$$

$$j = -\frac{m_o r_o c^2}{d^2}$$

---

[3]Because a particle and its anti particle have opposite spins, global angular momentum will be zero if particles and anti particles were created in pairs (ergo, the counter-rotational vortex of a quantized angular momentum need not balance a particle's angular momentum to zero).

ELECTRIC FIELD VS ANGULAR MOMENTUM SPACE

In **Part I**, we maturated Richard Feynman's *"pseudo force"* musings into an expansion resistant inertial force and called it gravity. Could the cosmological acceleration field also demystify the electric charge and its phenomenological peculiarities? Like gravity, the interaction between electric particles requires some form of continuous action. The template for our study will be the electron, properly it could be called " fundamental" as it prescribes the strength and polarity of the electric force as an extension of itself. Taking mass, angular momentum and expansion as the basic operatives, we formulaize a circulatory proto-particle and calculate the reaction.

Herein electron and positron mass is designated as m_0 (**9.1 x 10^{-31} kgm**), one electron unit of charge **q** equals (**1.6 x 10^{-19} coulombs**). Angular momentum $h/4\pi$ = (**5.3 x 10^{-35} kgm m^2/sec**) in any direction of measurement. The long range electric paradigm formulated in terms of mechanical properties (mass, angular momentum and vorticity), must account for attractive and repulsive forces **10^{42}g**. The ultimate question is whether charge is a elemental property of nature, in and of itself, or like Feynman's gravity, a flawed perception of something already known.

Our proto electron takes form as a spherical mass of radius 'r_0' (**1.41 x 10^{-15} meters**) wherein 'r_0' is determined by equating the energy m_0c^2 to that required to assemble a charge q^2 over the surface of a sphere. All electrons are alike, the quantized value of each parameter uniquely jig-sawed to complete the schematic. The challenge is that of fitting the pieces **q**, r_0, m_0 and $h/4\pi$ together to replicate the measured force between charged particles.[4] At the outset, a formulation of charge based solely upon the dimensional properties of electrons and positrons would seem to be in trouble. How can such a theory account for charged particles such as pions and protons and with different masses? Spin-less particles exhibiting charge and massless spin particles without charge, presents additional hurdles. At this early stage, the approach is definitely cavalier, the electron and positron are taken as definitive of charge. All that is known regarding the interaction between charged particles is characterized in terms of the clinical properties thereof. In anticipation of a resolution that will abrogate these obstacles, we proceed undaunted, leaving to the end a few paragraphs discussing the internal cancelling factors of compound particles and some thoughts of the electrical properties manifest by short lived massive particles such as pions and muons. That the muon, and other unstable massive particles, exhibit electric fields, can be adapted to the electron model without compromising the development of charge premised solely upon electron parameters, is admittedly, left partially unanswered by the present lack of knowledge as to their internal constituency. In such matters it is convenient to cop-out with a quote from Author Eddington:

> *"If we are not content with the dull accumulation of experimental facts, if we make*
> *any deductions or generalizations, if we seek for any theory to guide us, some degree*
> *of speculation cannot be avoided"*

---

[4]One fermi = **10^{-15}** meters. The electron radius r_0 is therefore 1.41 fermi (½ the classical electron radius r_e = 2.8 fermi)   To build the charge shell **q**, increments of infinitesimal charge are brought together from infinity. Since each new increment **dq** will experience a repulsive force, the energy $\int \mathbf{F(ds)}$ required to overcome the repulsive force to join the charge already assembled requires a total work energy $E = m_0 = K_eq^2/2r_0$ The resulting repulsive force $\mathbf{F_e}$ will be $K_eq^2/2(r_0)^2$. Substituting for K_eq^2 then r_0 = **1.41 fermi**

The inability of classical mechanics to explain quantum angular momentum in terms of rotating matter, like other peculiarities of subatomic physics, calls for an extension of the global acceleration epitome from force produced gravitational reactance to spatially expanding angular momentum. To extemporize an electron as circulatory space, we ideate a spherical central HUB of radius r_0, implicate mass m_0 and espouse the space/time ratio 'c' as the expansion driven operative.

Figure II-1A shows a vortical velocity profile for circulatory space wherein the velocity v_r decreases inversely with distance 'd' as would a physical fluid attracted by a central source. **Figure II-1B** shows virtual spatial flux having velocity 'c' at all radii. That the two boil down to one-in-the-same from an angular momentum perspective, will be underscored herein-below. As was the case with gravity, we begin by extending Newton's second law to relative acceleration. Taking relative circulatory velocity as 'c' between space and HUB corresponds to isotropic divergent space c^2/r_0. As with gravity, the action of global spatial expansion upon the rotational interface at r_0 creates a counter acceleration field that, when multiplied m_0 quantifies both the electrical reactive force $m_0 c^2/r_0$ and a yet to be explained local angular momentum field as $m_0 c r_0$. The actual form of m_0 is at this adjunct of our development depicted as a central sphere of unspecified radius. From another perspective, conservation of angular momentum during expansion requires $(d/dt)[(m_0)(v)(r)] = 0$, then:

$$\mathbf{dL/dt = [(m_0)v[dr/dt] + (m_0)r[dv/dt] + vr[d(m_0)/dt]] = 0} \qquad (2.1A)$$

Since m_0 is constant for present purposes, then $dr/dt = v = c$ at r_0 and therefore:

$$\mathbf{dv/dt = a = (-\, c^2/r_0)} \qquad (2.1B)$$

FIGURE II-1A   **FIGURE II-1B**

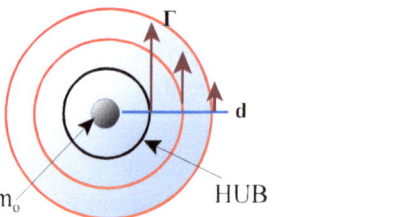 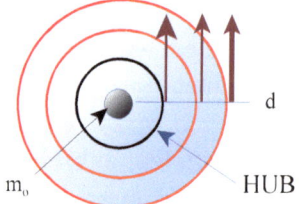

Figure II-1A illustrates the velocity profile for a free vortex (v·• r = constant). Fluid velocity increases as the radius diminishes reaching a maximum 'c' at radius r_0. In Figure II-1B, all spatial velocities are virtual and all have the same velocity 'c.' Circulatory force at d is c^2/d.

As with gravity, we look to the inertial dynamics of virtual motion to find a space-mass, role reversing, coordinate frame. Conceptually, the spatial circulatory field of electrons provides an ideal polar coordinate playing field. Centered upon the particle, and imposed with the communication condition that circulation at any distance transforms to the particle surface as $c^2/d \rightarrow v^2/r_0$, we debut our electron as pictured in **II-1B**. The circulatory velocity at any distance 'd' is 'c.' Each stream-tube of circulation at distance 'd' is inversely coupled (1/d) to the spherical energy core m_0 of radius r_0. (what will later be identified as the effective radius of reflexed circulatory energy of the field.

Symbolically, 'c' defines the iconic space/time relationship in dimensional units of velocity, but in the context of massless space, nothing moves. Here, as with gravity, we inculcate an inertial reactive field rationalized from second law symmetry.

The spatial circulation field of the electron is driven by spatial expansion. Reactionary centripetal acceleration diminishes inversely with distance 'd.' The acceleration at r_0 will be c^2/r_0, and c^2/R at the Hubble limit. Virtual centripetal force $m_0 c^2/r_0$ at the HUB thus corresponds to a local angular momentum $m_0 c r_0$ and the total spatial angular momentum field corresponds to $h/4\pi$. Consistent with what is developed herein, the properties of a free vortex are concentrated at its rotational center, and for virtual circulatory space, the effect upon angular momentum diminishes in the same ratio as vortical velocity in a material vortex. Circulatory space is mathematically analogous to frictionless flow along a stream-tube i.e., $(dv/v + dr/r) = 0$, whence $\mathbf{v} \cdot \mathbf{r}$ = Constant. However, in the case of the electron, 'r' cannot be zero, hence the velocity-distance product is constant, and therefore, circulation is the same at all distances 'd' greater than r_0:

$$\mathbf{(v) \times (d) = [c] \times [r_0]} \tag{2.1C}$$

The HUB mass m_0 then becomes a critical element. The inertial reactive property of m_0 must be concurrently coupled to all points of the 3-D spatial vorticity field, that is, all virtual circulations of any radii and in every rotational plane? Ergo, the HUB mass m_0 cannot be condensed matter in the traditional sense, nor can it be envisioned as simultaneous rotation about more than one axis. In **Figures II-1A** and **II-1B**, m_0 is depicted as a sphere coincident with the systemic center, but in reality no such chunk of matter exists, it is simply a measure of reflexed rotational energy The influence of spatial circulation upon m_0 at greater distances is reduced in accordance with (2.1C).[5] Effective centripetal force F_d at the HUB interface will be $(m_0)v^2/r_0$ and the isotropic acceleration force F_d between m_0 and space at any distance 'd' is then:

$$\mathbf{F_d = m_0 \left[\frac{v^2}{r_0} \right] = \left[\frac{m_0}{r_0} \right] \left[\frac{r_0 c}{d} \right]^2 = \frac{m_0 r_0 c^2}{d^2}} \tag{2.2}$$

Per (2.2), our proto-electron takes form as a virtual circulatory field having reflexed reactive mass m_0. Intensity falls off inverse square. The same follows from a theorem due to Stokes:[6]

$$\mathbf{\int_r v \cdot dl = \int\int_S n \cdot curl\ (v)\ dS} \tag{2.3}$$

For $\mathbf{v} = \mathbf{c}$, and $\int \mathbf{dl} = 2\pi\mathbf{d}$, the Curl $\mathbf{C}$ for a closed path of integration including the center is the line

_____

[5]A vortex made of nothing is the analog of divergent expansion of nothing. Both are virtual in the sense that the descriptions are mathematical, the universe behaves in a way that allows c^2/d vectors to be assigned to each point **P** of the field which correspond to the centripetal acceleration if local space where circulating about the HUB.

[6]Stokes Theorem declares the line integral of the velocity vector around a closed contour to be equal to the integral of the curl over the surface bounded by the contour

integral/Area :

$$2\pi cd/\pi d^2 = 2c/d \quad \{\text{for one rotational plane}\} \tag{2.4}$$

In a two dimensional plane surface, the integral around any closed curve that includes the center is $2\pi C$ The Curl $\{\nabla \times \mathbf{F}\}$ at a point $\mathbf{P}$ is the limit of the circulation per unit area (ΔS) at point $\mathbf{P}$ as the area ΔS shrinks to zero, expressed symbolically as:

$$\mathbf{n} \cdot (\nabla \times \mathbf{r}) = \frac{\lim}{\Delta S \to 0} \oint_C \mathbf{f} \cdot \mathbf{dr} \tag{2.5}$$

where $\mathbf{n}$ is the unit normal vector and ΔS is the area bounded by the contour over which the line integral is taken. If a closed path of integration does not include the center, Curl is zero. The rotational properties of a mathematic vortex are reflexive to its center, a dimension-less point $\mathbf{P}$ where velocity is infinite. **Figure II-1C** shows the constant 'c' circulatory spatial field of an electron and **II-1D** shows how the affect of spatial circulation diminishes with distance. The simulative treats the flux element of every stream-tube at any distance 'd' as having the same length **dl** (the distance traveled by an imaginary spatial element at velocity 'c' for a time increment Δt. Red Flux sweeps out a greater angle $d\theta$ than the brown, and the brown a greater angle than the green.

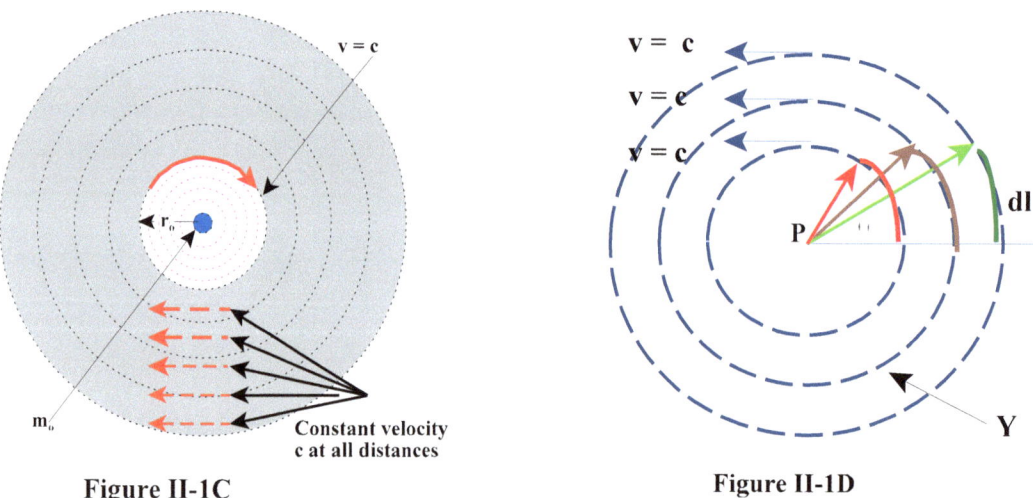

Figure II-1C   **Figure II-1D**

Figure II-1D. The red, green and brown radial markers illustrate the diminishing influence of circulatory spatial flux upon angular change $d\theta/dt = \omega$. Taking the path of integration over a spherical surface of radius 'd,' two orthogonal rotational planes will be contributing to centripetal force at each point on the surface of integration of area $4\pi d^2$. The effect of the two Curls ($4c/d$) upon the angular momentum about any axis is therefore c/d.

Comparison of the gravitational and electrical formulations in terms of inertial matter reveals similarities. The intensity of both diminishes inverse square. For gravity wrt distance from the reactive center. From **Part I**, The 'g' field of a uniform 3-sphere mass **M** of radius r_s at **d** is:

$$E_g = \frac{Mc^2}{R}\left[\frac{1}{4\pi d^2}\right]\left(\frac{1}{\sigma}\right) = \frac{Mc^2}{4\pi R(d^2)\,\sigma} \tag{2.6A}$$

whereas for an electron, the force $m_o c^2 / r_o$ spread over a spherical surface of area $4\pi d^2$ falls off in accordance with $4\pi r_o^2/4\pi d^2$

$$F_e = \frac{m_o c^2}{r_o}\left[\frac{4\pi r_o^2}{4\pi d^2}\right] = \frac{c^2 r_o m_o}{d^2} \tag{2.6B}$$

Both specify force in terms of mass energy (mc^2) divided by distance squared.

From the standpoint of action-reaction, the mechanical model must take into account three planes of spatial rotation. However, to calculate the effect of the field at a particular point it is only necessary to consider two spin planes, that is, as between two points separated in space, only two orthogonal planes can intersect both points. If the mechanical model is to merit further study, it should now be tested by comparing the predicted normal force F_n at r_o (equation 2.4) against the self repulsive coulomb force created by a charge **q** distributed over a spherical surface of radius r_o. The electrical force is F_e obtains by merging two charges together unto the shell and dividing by **2**, that is:

$$F_e = \frac{k_e(q_e)^2}{2(r_o)^2} = \frac{(9\text{x}10^9\,\text{kgm}\cdot\text{m}^2/\text{coul}^2)(1.6\text{x}10^{-9}\,\text{coul})^2}{2r_o^2}\cdot = \frac{11.52\,\text{x}10^{-29}\,\text{ntn}\cdot\text{m}^2}{r_o^2} \tag{2.6C}$$

For the mechanical model:

$$F_n = \frac{m_o r_o c^2}{r_o^2} = \frac{(9.1\text{x}10^{-31}\,\text{kgm})(1.4\text{x}10^{-15}\,\text{m})(3\text{x}10^8\,\text{m}/\text{sec})^2}{r_o^2} = \frac{11.46\text{x}10^{-29}\,\text{ntn m}^2}{r_o^2} \tag{2..6D}$$

Within the accuracy of the parameters used for the comparison, the two expressions define the same force. All electrons are alike because there is only one combination of reflexed energy, angular momentum and spatial circulation that can exist as a stable configuration. As later shown (2-15 and 2-16), the mass energy of the electric field must correspond to the HUB energy which determines the HUB radius which in turn defines the value of the electric field in terms of c^2. And also as to be later shown, it is the electric field that determines the spatial angular momentum field, or more properly, they are one in the same.

Figure II-2 shows two identical particle systems β_1 and β_2 each comprising a mass $\mathbf{m_0}$ and spatial angular momentum $\mathbf{L_s}$ characterized as circulatory virtual flow in accordance with (2.2) in two dimensional **X-Y** space—it being understood that in the absence of an aligning field, the angular momentum spin axis may assume any and all angles with respect to an arbitrary coordinate system[7] Both particles are assigned clockwise rotations in two dimensional space. The separation distance $\mathbf{d} >> \mathbf{r_0}$. The two parallel lines $\mathbf{Y_1}$-$\mathbf{Y_1}$ and $\mathbf{Y_2}$-$\mathbf{Y_2}$ are drawn through the centers of the particles so as to divide the **X-Y** space into three parts. In the absence of influence by the other, the spin field of each system will be spherically symmetrical with respect to mass $\mathbf{m_0}$. In the presence of the other particle, each field is unbalanced. The spin field of β_2 is opposite to the spin field of β_1 between $\mathbf{Y_1}$ and $\mathbf{Y_2}$ and additive in the region above $\mathbf{Y_1}$. Similarly, the spin field of β_1 counteracts the spin field of β_2 between $\mathbf{Y_1}$ and $\mathbf{Y_2}$ and augments it in the region below $\mathbf{Y_2}$. This superposition of spins carries a consequence: When the fields are additive, centripetal force acting between $\mathbf{m_0}$ and space will be greater than that due to its own circular field. When the field of the other opposes the circulation, net centripetal force is decreased. As a result, the central mass $\mathbf{m_0}$ of each particle system will be subjected to greater force in one direction than the other. The net force upon the masses $\mathbf{m_0}$ of β_1 and β_2 will be oppositely directed. Like charges repel[8]

---

[7]The superposition of simultaneous spatial vortices to define a spherical surface does not create a problem in a massless medium—there being no traffic congestion in the crossing paths of massless virtual fluids. In actuality, the particles are three dimensional and spatial rotations about any axis is quantized and uncertain. In a *"field free"* environment, the precepts of quantum mechanics certify the spin plane of each vortex remain undefined until orientation is determined by measurement which typically involves the presence of other particles, in which case rotations of like particles orient to be repulsive per **Figure 11.**

[8]The mathematical underpinning for the universe does not demand that space be physically definable in terms of substantive properties. It does have operative properties by which energy, force and motion are rendered null on a global scale. For electrons and positrons, expansion of the spatial angular momentum field sustains the virtual mass $\mathbf{m_0}$ in balance with that necessary to create the electric field, which defines the effective radius $\mathbf{r_0}$ and consequently the electrical force $\mathbf{m_0 c^2/r_0}$ that in turn exerts the reactionary convergence necessary to retain the virtual vortical field $\mathbf{m_0 c^2 r_0/d^2}$ at all distances 'd.' Electrons and positrons can only exist as self sustaining organisms in a spatially expanding environment.

FIGURE II-2

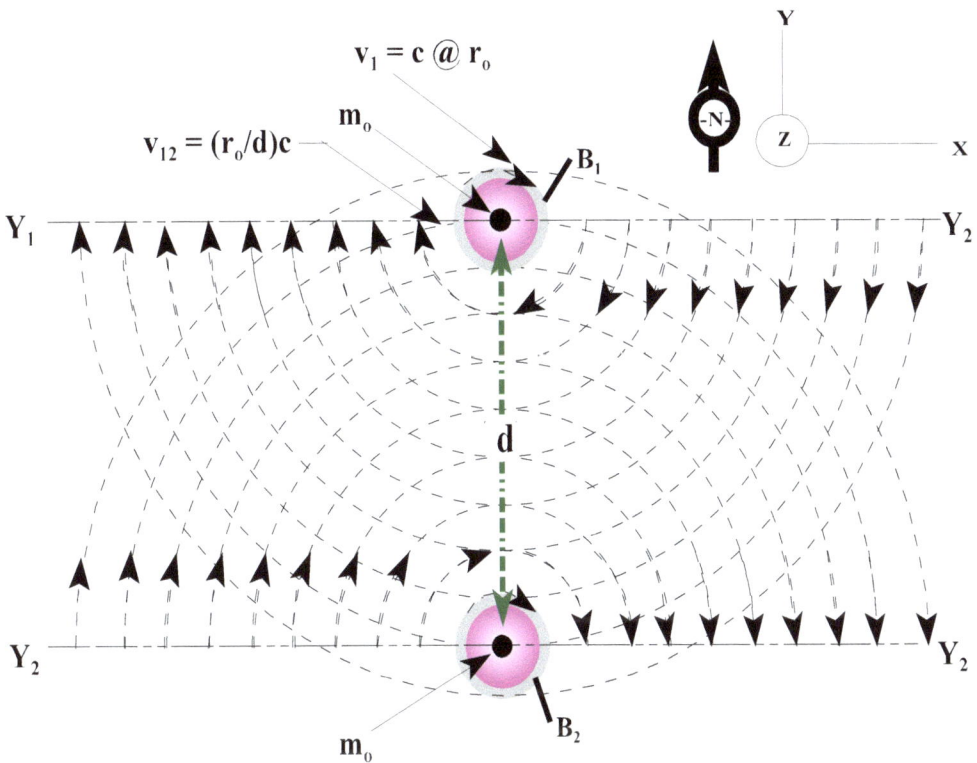

Figure II-2. **The coupling between circulatory fields is shown as being simultaneously both additive and subtractive. The two virtual spatial rotations in the x-y plane each have peripheral spin velocity "c" at radius r_o, and each encompasses an identical symbolic mass m_o. Both circulations are clockwise in the X/Y plane which is divided into three areas by the two parallel east-west lines Y_1-Y_1 and Y_2-Y_2. In the hinter region between these lines the vortical field of each particle counteracts the local angular momentum of the other. In the space north of Y_1-Y_1 the field of β_2 is depicted as augmenting the vortical strength of β_1 and in the area south of Y_2-Y_2 the field of β_1 is depicted as bolstering the strength of β_2.[9] Superposition of the two fields results in an unbalance in the force exerted upon each of the masses m_o. The net force upon β_2 will be southward, and that upon B_1 will be northward. Like charges repel.**

---

[9]The influence of β_2 on β_1 in the area North of y_1-y_1 and β_1 on β_2 in the area south of y_2-y_2 would appear to be limited to the velocity of light c. Since the velocity at the radius ro is 'c' even in the absence of the other particle, the effectiveness of the velocity boosting component of each field may be a nullity. In that case the total force per rotational plane is half that calculated in (2.11). However, there are always two equally effective orthogonal circulations affecting the force acting upon the masses m_o of each system.

The Force Between Charges

Current **QED** theory purports to explain electrical phenomena in terms of virtual photons supposedly brought into existence by the intensity of the very fields sought to be explained. The theory itself does not predict the Coulomb force from the known properties of electrons and photons, rather, it owes its endorsement to an extremely accurate prediction of the anomalous *Gyromagnetic* spin ratio.[10] **QED** takes into account the wave nature of the electron, together with the premise that particles only transmit force in increments—the coupling between physical entities being adjusted along the way to correspond to the probability of certain occurrences identified with the strength of the perturbation. The predicted results, however, relate only to a second order effect, and not to a quantitative expression for the force.[11] What we seek here is a physical theory which explains the strength of the electron charge and how it arises. To calculate the repulsion between electrons in terms of first principles, we must take a different path.

Figure II-3 shows a vortical particle system β_1 within the influence of a counterclockwise rotational field of like origin (particle β_2 not shown). The distance **d** separating β_1 and β_2 is large in comparison with r_o which permits representation of the (v_{12}) field produced by β_2 as equally spaced straight lines orthogonal to the line of action drawn between the radial centers of the two particles β_1 and β_2. What is desired is an expression for the combined velocity field at all points of superposition on the surface defined by the radius r_o. From this, we will calculate the centripetal force exerted by the effective velocity at all points. The sum of the components of these forces resolved along the line connecting the particle centers is the objective.

For the particle β_1, the force produced by the superposition of the two velocity fields in the northern (top) hemisphere is given by the square of the sum of the β_1 field (v_1) plus the component of the β_2 field (v_{12}) which is parallel to the (v_1) field, that is:

$$(M_o/r)(v_1 + v_{12} \sin \theta)^2 \qquad (2.7)$$

and in the southern hemisphere, the force is given by the square of the difference between (v_1) and the (v_{12}) component parallel to (v_1) i.e.,

$$(M_o/r)(v_1 - v_{12} \sin \theta)^2 \qquad (2\text{-}8)$$

where θ is the angle measured from the east-west line Y_1-Y_2 which bisects β_1.

Net radial force F_r exerted by the superposition of the velocity fields at any two points

---

[10] In **Q**uantum-**E**lectro-**D**ynamical calculations, electrons are described by wavefunctions that exists throughout space. To find the force between two such particles, one must calculate all the probabilities represented by the squared amplitudes of the waves at each point, and then add up the results. Since the wavefunctions overlap at certain places, there will be some locations where the force contributions are infinite. These infinities are disposed by a mathematical contrivance dubbed re-normalization—after which **QED** gives the right values for the anomalous magnetogyric ratio.

[11] **QED** should be applauded for what it is and what it does predict, but it is not a proper theory for explaining the origin of the Coulomb force, which it does not do.

intersected by the north-south meridian line X is equal to the difference between the squares of the

$$F_r = \left(\frac{m_o}{r_o}\right)\left[\left(v_1 + v_{12}\sin\theta\right)^2 - \left(v_1 - v_{12}\sin\theta\right)^2\right] \quad \text{velocities, i.e.,}$$

$$= \left(\frac{m_o}{r_o}\right)\left(4v_1\right)\left(v_{12}\sin\theta\right) \tag{2.9}$$

The component F_x of the radial force F_r along the line of action joining the two particles is:

$$F_x = \left(\frac{m_o}{r_o}\right)\left(4v_1\right)\left(v_{12}\sin\theta\right)\left(\sin\theta\right) \tag{2-10}$$

The average value F_A of the north-south component of the unbalancing force as the angle θ varies from zero to π is:

$$F_A = \frac{4m_o v_1 v_{12}}{\pi r_o} \int_0^\pi \sin^2\theta \, d\theta$$

$$= \frac{4m_o v_1 v_{12}}{\pi r_o} \left[\frac{\theta}{2} - \frac{\sin 2\theta}{4}\right]_0^\pi \tag{2-11}$$

$$= \left(2m_o/r_o\right)\left(v_1 v_{12}\right)$$

Comparison of equations (2-4) and (2-11) suggests we have arrived at the correct force for the operative action in both planes. If c is the peripheral velocity v_1 at r_o and $c(r_o/d)^2$ is the velocity field v_{12} at distance d, then the correct force F would be double (2.12).

$$F = \frac{2m_o c^2 r_o}{d^2} \tag{2-12}$$

Figure II-3. A two dimensional particle system β_1 having counterclockwise rotation is separated from an identical particle β_2 (not shown) by a distance 'd' which is large in comparison with the nominal radius r_o. The local velocity field of β_2 can therefore be approximated as uniformly spaced straight lines v_{12} from east to west. The components of v_{12} that are to be added or subtracted from the velocity v_1 is given by the sin of the angle θ between v_1 and v_{12}. Each point at an angle θ in the northern hemisphere corresponds to a point at an angle $-\theta$ in the southern hemisphere. These complimentary locations are conjugate, the velocity v_{12} being additive in the northern hemisphere and subtractive in the southern.[12] Note shown are the interacting spatial circulations in the plane normal to the page which also passes through the centers of β_1 and β_2.

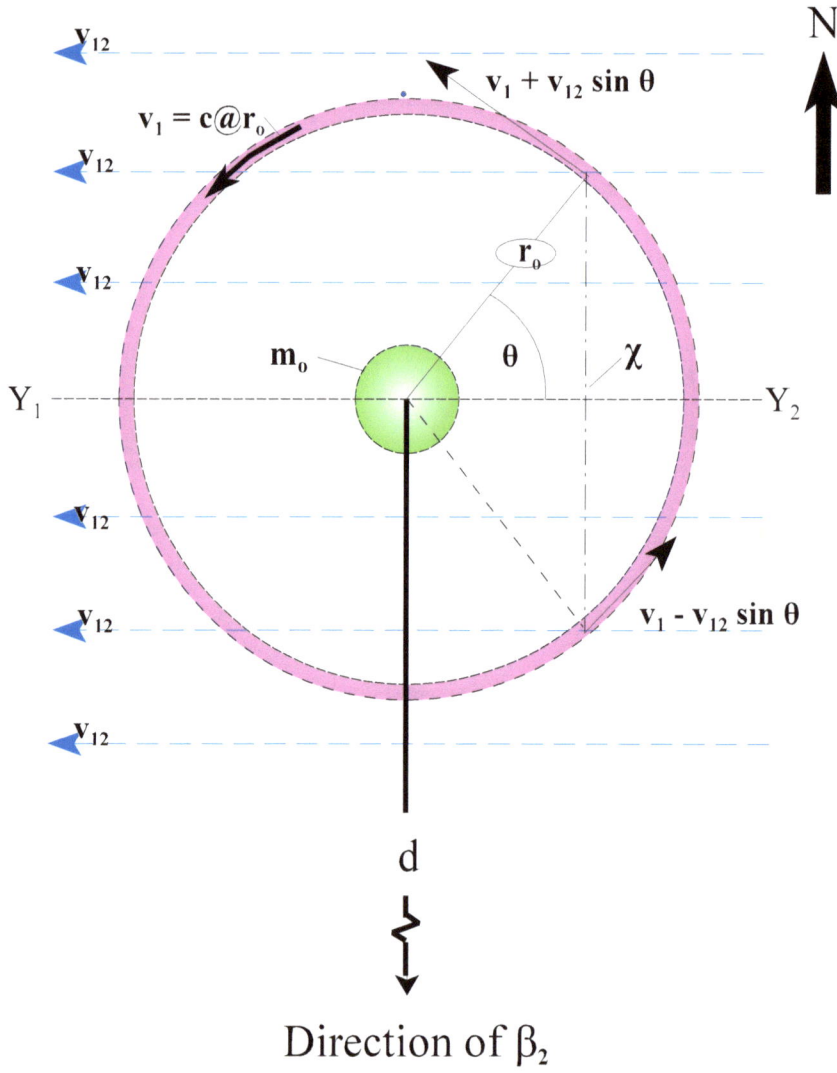

Direction of β_2

[12] If $(v_{12} + v_1)$ exceeds "c" the analysis must be modified as per pages 62-64 infra.

We have arrived at the correct numerical result without considering circulatory interaction in the orthogonal plane. Three dimensional spatial circulation reduces to interaction in the two orthogonal planes that pass through the centers of both particles. No component of rotation in a third plane perpendicular to the two orthogonal planes that pass through the electron centers can effect a force upon the electrons. Since (2-12) correctly predicts the Coulomb force as vortical interaction between two rotational fields, we conclude that superposition of one rotational field upon the other is only effective to reduce the field of the other, i.e., circulatory velocity cannot be augmented to exceed 'c' for any region of space. With this correction (based upon c velocity at all distances) we accept (2-12) as correct in that it overstates the force in one plane by 50% but fails to account for interaction in the orthogonal plane which contributes ½ of the composite force.

If either one of the circulations is reversed, the direction of force will be likewise reversed. Suffice it to say, the force on β_2 will always be equal and opposite to that of β_1 since each produces an identical influence upon the field of the other, a commandment of Newton's 3rd law. [13]

Coulomb's Law vs Spatial Vortex

$$\mathbf{F_e} = \mathbf{k_e} \frac{\mathbf{q_e}^2}{\mathbf{d^2}}$$

$$F_A = \frac{2m_o c^2 (r_o)}{d^2}$$

$$= \frac{\mathbf{2.3x10^{-28} ntn \cdot m^2}}{\mathbf{d^2}}$$

$$= \frac{2.3x10^{-28} ntn \cdot m^2}{d^2}$$

Minimization of energy during interaction is a systemic stipulation of interacting fields, whether it be moving space or moving mass. Since less energy is required to imbalance the near field rather than reinforce the far field, the spin planes orient so that two of the three will occupy a state of opposition, consequently, the effective velocity field in the hinter region between the particles is diminished. Lower effective velocity decreases overall energy of the two particle system. The total energy of the field coupled structure is minimized when the in-between velocity field is minimized. The specification of electrical charge in terms of first principles is thus:

$$\mathbf{E_e} = \frac{\mathbf{k_e (q_e)^2}}{\mathbf{2r_o}} \tag{2.13}$$

And since $\mathbf{E_e} = \mathbf{m_o c^2}$, then:

$$\mathbf{k_e q_e^2 = 2m_o c^2 (r_o)} \tag{2.14}$$

Equation (2-14) is the electro-mechanical transform between the two divisions of classical physics. To shift from one to the other, substitute for $(\mathbf{k_e q_e^2})$ in Coulomb's law, or $(\mathbf{2m_o c^2 r_o})$ in the mechanical formulation. From (2.17) below, the energy in the electric field is $(- \mathbf{m_o c^2})$.

---

[13] For $q_e = 1.6 \times 10^{-19}$ coul, $r_o = 1.4 \times 10^{-15}$ meters, $m_o = 9.1 \times 10^{-31}$ kgm, and $k_e = 9 \times 10^9$ kgm m²/coul²

The ultimate question of charge as a fundamental property of nature must be rethought. In its fundamental form as a bare electron, spatial vorticity is defined by Newtonian reaction. If the formulation is correct, the virtual circulatory field summed over the volume of the Hubble sphere must correspond to the angular momentum quantum $h/4\pi$.

Conservation of angular momentum is not overruled by expansion. As per (2.14) the dilation of an angular momentum space will be accompanied by an inwardly directed reactive force, that which gives rise to the electric force field. To express the energy contained in an expanding volume of empty space, we revert to the electrical formalism. Comparison of the kinetic energy of circulatory space contained in the expanding Hubble volume to the positive energy of the non-expanding hub of radius r_0, shows that the two are equal and opposite.[14] The energy of the electric charge is the integral over the energy density:

$$\frac{\varepsilon_0}{2} \int_V \mathbf{E} \cdot \mathbf{E}(\mathbf{dV}) = \int_{r=0}^{R} \frac{q^2}{32\pi^2 \varepsilon_0 r^4} = -\frac{q^2}{8\pi\varepsilon_0} \frac{1}{r} \tag{2.15}$$

Using the electrical relationship $4\pi\varepsilon_0 = 1/k_e$ then from (2-14):

$$\mathbf{Energy = 2m_0c^2(r_0)/2r \big|^{r=R} - 2m_0c^2(r_0)/2r \big|_{r=0}} \tag{2.16}$$

which straightaway revives the age-old problem of particles as points. Clearly, there is no difficulty with the limit $r = R$, but for $r = $ zero, the field energy per (2.16) is infinite. The difficulty evaporates, however, if the lower limit $r = r_0$ which comports with our objective in that (2.16) reduces to:

$$\mathbf{Field\ Energy = (-\ m_0c^2)} \tag{2.17}$$

As initially prefaced, the electrical and mechanical formalisms (2.14) are based upon empirical quantities. Accordingly, electron mass-energy for present purposes is a quantified constant uniquely interdependent upon r_0. Total energy in the circulatory prescription is also $\mathbf{m_0c^2}$ per (2.17), $\mathbf{r_0}$ being correspondingly fixed and quantified consequent thereto.

In the Inertial-Gravity confluence related in **Part I**, negative gravitational energy and positive mass energy were axiomatically balanced. For the electron, negative circulatory energy is reflected to the HUB in form as positive energy $\mathbf{m_0}$. When virtual circulatory flow is treated as uniform (constant velocity $\mathbf{c}$ at all distances), the contribution of the flow to angular momentum (at any distance '$\mathbf{d}$' from the rotational center) is distributed over the spatial length $\mathbf{2\pi d.}$ As previously belabored, the effect of circulation diminishes inversely with '$\mathbf{d}$,' per (2.4), the centripetal force also diminishes as '$\mathbf{1/d}$' per (2.1A), and the field force diminishes as $\mathbf{1/d^2}$ per (2.2).

---

[14]Equal and opposite circulations are presumed to arise during the initial instant of expansion when stress intensity was maximum. There is never net energy due to the consequent $\mathbf{G}$ field and never a net global angular momentum since the number of positrons locked up in protons will be always equal to the number of electrons .

So while circulatory space properly predicts repulsive and attractive forces between like and unlike charges, respectively, it also explains Feynman's enigma, the irreconcilable isotropic angular momentum quantum **h/4π**. The electron angular momentum **h/4π** cannot be local rotational mass, but it can be 3-D virtual circulatory flux reflexed to the HUB.[15]

Since the orientation of circulatory spatial flux is not limited to a single plane, when measured as a property of the particle, it will exhibit no directional preference. The challenge is that of showing the electric field to be the alter-identity of the spatial angular momentum field **h/4π**.

To complete the model, the electric field will be expressed in terms of the spatial angular momentum field. The claim is then made that the sum total of the energy in the spatial rotational field is m_oc^2 and that the spatial circulation corresponds to an isotropic angular momentum **h/4π**.

The circulation at distance 'd' from the rotational center of a plane passing through the rotational center is from (2.1C), cr_o/d. Just as there are two orthogonal planes of spatial rotation contributing to the electric force, there are two orthogonal rotational planes intersecting both the center of rotation and an arbitrary point in space. They add vectorially as √2. Spatial angular momentum is therefore the square root of the sum of the squares of two orthogonal circulations multiplied by the central mass m_o. The sum of the moment of momentum LC_T of two intersecting orthogonal circulation fields is:

$$\mathbf{LC_T} = \sqrt{2}(\mathbf{m_o})\int_{r_o}^{R}\frac{\mathbf{cr_o}}{\mathbf{r}}(\mathbf{dr}) = \sqrt{(2)}\mathbf{cr_o}(\mathbf{lnR} - \mathbf{lnr_o})(\mathbf{m_o})$$

$$\approx \sqrt{2}(\mathbf{cr_o})(\mathbf{ln}10^{26} - \mathbf{ln}10^{-15})(\mathbf{m_o})$$

$$\approx (1.414)(3\mathrm{x}10^8)(1.4\mathrm{x}10^{-15})(60 - [-35])(9.1\mathrm{x}10^{-31})$$

$$\approx 5.1\mathrm{x}10^{-35}\,\mathbf{meters^2kgm \cdot sec^{-1}} \approx \mathbf{h}\!\!\!\Big/\!_{\mathbf{4}\pi}$$

(2.18)

This completes our composition of the electron (or positron) as virtual circulatory space, to be understood as the multiplicatus angular momentum function **h/4π** of every subatomic particle which, although not always measurable because of cancelling counter oriented spins, or measured in some multiple of **h/4π** in complex particles composed of more fundamental entities. Postulation is superseded by virtual circulation, the electron now exposed as a charade where expanding spatial angular momentum poses as the electric field. No part of the electron complex is independent. Taken together with the space/time operative '**c**,' the quantified values of mass m_o, angular momentum **ħ/2** and radius r_o form a fundament called the electric unit of charge **q**. Indirectly, we have raised the question of whether all mass can be explained in terms of reflexed spatial vorticity.[16]

---

[15]For circulatory flow, the integral around any closed path which includes the rotational center is **2πrv**, the circulation **C** will be **2πC** for all paths which include the center and zero for all contours which exclude the center.

[16]The idea of mass as degenerate spatial motion comports with its two long range fields, one a consequence of reactance imposed by m_o in opposition to spatial divergence, the other the reactance created by m_o as the constellation of the reactive energy created by expanding spatial vorticity.

ALPHA

The ratio of the electron velocity in the first Bohr orbit to the speed of light is symbolized by the first letter of the Greek alphabet, α. Also known as the Sommerfield fine structure constant, "*alpha*" comes into play whenever waves interact with particles.[17] Its symbolic value $k_e q^2 / \hbar c$ (approximately 1/137) has always been a mystery (why should these apparently unrelated constants, combine to create a dimension-less number [approx (1/137)]?

The "mechanical formulation" of the electron provides a Rosetta Stone. When expressed in terms of mass and size, "*alpha*" demystifies as the ratio of two angular momentums, namely the Hub angular momentum $m_0 cr$ divided by the angular momentum of the virtual circulatory field $h/4\pi$. Specifically, from (2.14), we recognize $k_e q^2$ as $2m_0 c^2 r_0$. Rewriting alpha as $[k_e q^2/c]/\hbar$, then

$$\alpha = k_e q^2 / \hbar c = [k_e q^2/c]/\hbar = 2m_0 c^2 r_0 / (ch/2\pi) \tag{2.19}$$

which reduces to:

$$\alpha = \frac{m_0 cr_0}{\dfrac{h}{4\pi}} \tag{2.20}$$

Expressed in terms of its mechanical pedigree, alpha reveals the action of the whole in terms of its parts. Negative field energy (2.15) reflexed to the HUB ghosts as the positive rotational mass energy $m_0 c^2$ that defines the virtual Hub angular momentum $m_0 cr_0$. The expansion generated centripetal force $m_0 c^2/d$ thus comports with the acceleration at distance '**d**' multiplied by m_0. Specifically, if $m_0 c^2/r_0$ is the force at the Hub, then:

$$F_{HUB} = (m_0 c^2/r_0)(r_0/r_0) \tag{2.21}$$

is also the force at the Hub, which can be written as:

$$F_{HUB} = m_0 c^2 r_0 / (r_0)^2 \tag{2.22}$$

Ergo, for any other distance '**d**' the force due to the rotational interaction will be:

$$F_d = m_0 c^2 r_0 / (d)^2 \tag{2.23}$$

which expresses the continuity of the electron state as one in the same with (2.2). In words, the inverse dependance of centripetal force upon distance assures the continuity of '**c**' circulatory formalism at all distances). And since HUB angular momentum is simply '**c**' velocity rotation of the electric field energy m_0 reflexed thereto, the whole can now be understood as a self maintaining composition.

---

[17]The constant α plays a crucial role in **QED** theory in that it relates the electrostatic quantities k_e and **q** to the **h** and **c**

The fundamental nature of electric charge has been reduced to a ratio between mechanical and spatial circulation thus posing two issues, 1) that of why some spin particles exhibit no electric field and 2) that of why some spin-less particles manifest the same electric intensity as electrons.

The first question is exemplified by neutrinos and photons. Both are fundamental entities. The combination of factors used to construct the mechanical model of the electron would predict neutrinos as charge-less, as is the case, since they have no ponderable mass upon which acceleration fields can act. The same is true of photons, moreover, photon angular momentum is coaxial and therefore does not create 3-D circulatory fields, photons should not manifest electric charge, as is also the case. Neutrons and protons are complex particles, and like all complex structures, net charge will depend upon the ingredients.[18]

The existence of spin-less electric particles with large masses, however, would seem to discredit any theory of charge based upon spin angular momentum. The deciphering of "alpha" suggests a rationale. For the electron, the mass and size is insufficient to explain the magnitude of its directionally resolved angular momentum $h/4\pi$. Local angular momentum is limited to the reflexed mass-energy m_o, the maximum possible HUB angular momentum is therefore $m_o c r_o$. For a heavy electron (symbolized as μ^- and called the mu minus meson), the mass factor is $(206.78)m_o$. The reason why particles such as $\pi+$ and π^- mesons exhibit zero angular momentum is because charged pions are complex entities composed of one neutrino and one muon, the latter determining pion polarity.[19] Analogously formulated using the muonic dimensional factors m_u and r_u in place of m_o and r_o, local angular momentum $m_u c r_u$ will cancel neutrino angular momentum when local angular momentum equals the circulatory electric field angular momentum $h/4\pi$, that is:[20]

$$F_u = \frac{m_u c^2 r_u \alpha}{d^2}$$

where (2.24)

$$r_u = \frac{h}{4\pi c m_u} = 0.95 \times (10)^{-15} \text{ meters}$$

---

[18] In the "standard theory" the constituents of protons and neutrons called quarks are hypothesized as fractional charges ($1/3q$, $2/3q$), but physically separated fractional spin entities have yet to be directly observed. While cancellation of spins in one or two dimensions of a 3-space circulation may be interpreted as fractionally charged separate particles, they can also be considered as parts of the whole. Indeed some theorists prefer the name partons as more representative of the physiology. The "standard model" is intended primarily as a mathematical artifice for predicting behavior rather than as a physical emulative.

[19]There is also a Tau particle with mass in the neighborhood of 3000 m_o also considered a fundamental, but like the muon, a neutrino (the tau neutrino) is released during decay. The above speculations regarding the true nature of the muon also apply the tau.

[20]The analogy here is that of two counter rotating wheels on the same shaft, one inside the other. If each flywheel exhibits the same angular momentum - net angular momentum is zero, but the larger wheel is capable of exerting an influence upon its surroundings at a greater distance. The analogy of the other flywheel as circulatory space would of course need to be revised to reflect the constant flux velocity 'c' at all radii.

If charged π mesons are a door to understanding quantum structure in terms of classical mechanics, then alpha is the key. There are then two formulations for **q**, one based upon the electron's radius and mass as determined by the circulatory field energy $\mathbf{m_o}$, the other based upon the properties of the muon. The ratio of angular momentum derived from the reflected rotational field energy to the angular momentum field is *alpha*. In the muonic formulation, increased mass is reflected in smaller radius determined within the containment structure. Muon mass being sufficient to cancel angular momentum locally, results in a condition of local angular momentum equal to global angular momentum, $\mathbf{m_u c r_u = h/4\pi}$. Upon escape from the internal force constrains imposed by the bound state of a muon as part of a complex particle, the muon decays into an electron and two hypothesized particles from the neutrino family needed to balance energy and angular momentum continuity. That charged pions decay to a muon of the same charge plus one neutrino is consistent with the existence of algebraically subtractive local spins. The sequence of events from pi-plus to mu-plus to positron is shown in **Figure II-4**.[21]

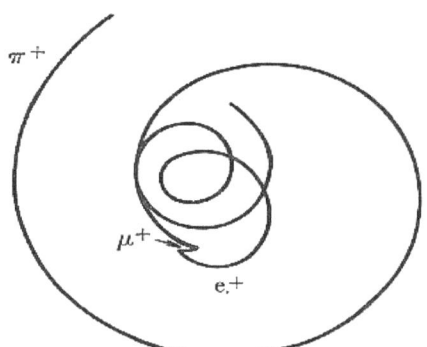

Figure II-4 shows the decay of a (pi-plus) to (mu-plus) and neutrino (the latter theorized, but not observed). (mu-plus) then decays to a positron with the emission of a neutrino and anti neutrino (both theorized but not observed). [From liquid hydrogen bubble chamber, Brookhaven National Laboratory]

The existence of rival particles manifesting the same quantum of charge as electrons whose properties formulaize as a unitized quantum **q**, is initially bothersome. But if the electron is viewed as the free state of a mu minus, the unyoked charge simply expands to its dimensional maximum, namely, the radius $\mathbf{r_o}$ corresponding approximately to the electrical energy $\mathbf{m_o}$ contained in the electric field of the muon prior to transition, i.e., the circulatory energy of the muon spatial angular momentum $\mathbf{h/4\pi}$, aka its electric field. While the tau and muon are regarded as fundamental particles, they come into free existence as an altered state of what soon disintegrates into an electron neutrino and anti-neutrino. That muons and taus can be internally stressed forms of matter that counter balance spatial angular momentum in a bound state as charged pi meson is consistent with what zero net annular momentum. Additional mass corresponds to the binding energy of two neutrinos, a smaller size and greater internal stress in the bound form.[22] While the muon is considered a fundamental particle having no place in the standard model, it is never alone–being either internally locked with other subatomic entities or a temporarily state defined by neutrinos.

---

[21] In approximately 1% of mu-minus decays, the pion transforms directly to an electron and anti neutrino

[22] In standard theory, the muon has been neglected entirely - there seems to be no place for it to fit

In classical physics, the idea of force per unit area, is paramount. In quantum mechanics, energy and momentum are the basic descriptors. The present study reverts to the ascription of spatial dimensions to subatomic entities. Most of these lengths can only be determined inferentially by observing other characteristics consistent therewith. What is significant from the perspective of the electron is that its local properties are determined by the extent of its global circulatory field. One might imagine electrons as muons freed from compactification stress. When bound with other particles, the electric field is suppressed, mass accordingly elevated. In these speculative(s), we are reminded of an idea long cherished by John Wheeler:

"Perhaps a different theory would reveal how all matter could be made from electrons."

And then perhaps, all matter can be traced to spatial vorticity.

Copyright B. Jimerson, January 2016

Interested readers are invited to send comments to "Cosmodynamics@yahoo.com"